180 Days of GEOGRAPHY
for Third Grade

Author

Saskia Lacey

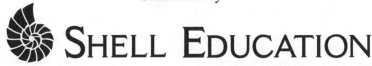

SHELL EDUCATION

Series Consultant

Nicholas Baker, Ed.D.
Supervisor of Curriculum and Instruction
Colonial School District, DE

Publishing Credits

Corinne Burton, M.A.Ed., *Publisher*
Conni Medina, M.A.Ed., *Managing Editor*
Emily R. Smith, M.A.Ed., *Content Director*
Veronique Bos, *Creative Director*
Shaun N. Bernadou, *Art Director*
Lynette Ordoñez, *Editor*
Kevin Pham, *Graphic Designer*
Stephanie Bernard, *Associate Editor*

Image Credits

p.39 British Library Board/Bridgeman Images; p.61 Library of Congress [LC-H2- B-3470-1]; p.75 Granger Academic; p.103 Joseph Sohm/Shutterstock; p.113 Michel Mond/Shutterstock; p.129 Reuters/Daniel Beltra-Greenpeace/Newscom; p.145 Library of Congress [LC-USZ62-63648]; p.167 Bridgeman Images; p.177 Library of Congress [LC-USZ62-132969]; p.187 Mark Wilson/Getty Images; p.211 Joyce M. Ranieri/Flickr; p.221 Library of Congress [LC-DIG-pga-01767]; p.228 Library of Congress [LC-USZ62-7823]; p.229 Library of Congress [LC-DIG-pga-03004]; p.231 Library of Congress [LC-DIG-ppmsca-04297]; all other images from iStock and/or Shutterstock.

Standards

© 2012 National Council for Geographic Education
© 2014 Mid-continent Research for Education and Learning (McREL)

For information on how this resource meets national and other state standards, see pages 10–14. You may also review this information by visiting our website at www.teachercreatedmaterials.com/administrators/correlations/ and following the on-screen directions.

Shell Education

A division of Teacher Created Materials
5301 Oceanus Drive
Huntington Beach, CA 92649-1030
www.tcmpub.com/shell-education

ISBN 978-1-4258-3304-6
©2018 Shell Educational Publishing, Inc.

TABLE OF CONTENTS

INTRODUCTION

With today's geographic technology, the world seems smaller than ever. Satellites can accurately measure the distance between any two points on the planet and give detailed instructions about how to get there in real time. This may lead some people to wonder why we still study geography.

While technology is helpful, it isn't always accurate. We may need to find detours around construction, use a trail map, outsmart our technology, and even be the creators of the next navigational technology.

But geography is also the study of cultures and how people interact with the physical world. People change the environment, and the environment affects how people live. People divide the land for a variety of reasons. Yet no matter how it is divided or why, people are at the heart of these decisions. To be responsible and civically engaged, students must learn to think in geographical terms.

The Need for Practice

To be successful in geography, students must understand how the physical world affects humanity. They must not only master map skills but also learn how to look at the world through a geographical lens. Through repeated practice, students will learn how a variety of factors affect the world in which they live.

Understanding Assessment

In addition to providing opportunities for frequent practice, teachers must be able to assess students' geographical understandings. This allows teachers to adequately address students' misconceptions, build on their current understandings, and challenge them appropriately. Assessment is a long-term process that involves careful analysis of student responses from a discussion, project, practice sheet, or test. The data gathered from assessments should be used to inform instruction: slow down, speed up, or reteach. This type of assessment is called *formative assessment*.

HOW TO USE THIS BOOK

Weekly Structure

The first two weeks of the book focus on map skills. By introducing these skills early in the year, students will have a strong foundation on which to build throughout the year. Each of the remaining 34 weeks will follow a regular weekly structure.

Each week, students will study a grade-level geography topic and a location in North America. Locations may be a town, a state, a region, or the whole continent.

Days 1 and 2 of each week focus on map skills. Days 3 and 4 allow students to apply information and data to what they have learned. Day 5 helps students connect what they have learned to themselves.

 Day 1—Reading Maps: Students will study a grade-appropriate map and answer questions about it.

 Day 2—Creating Maps: Students will create maps or add to an existing map.

 Day 3—Read About It: Students will read a text related to the topic or location for the week and answer text-dependent or photo-dependent questions about it.

 Day 4—Think About It: Students will analyze a chart, diagram, or other graphic related to the topic or location for the week and answer questions about it.

 Day 5—Geography and Me: Students will do an activity to connect what they learned to themselves.

Five Themes of Geography

Good geography teaching encompasses all five themes of geography: location, place, human-environment interaction, movement, and region. Location refers to the absolute and relative locations of a specific point or place. The place theme refers to the physical and human characteristics of a place. Human-environment interaction describes how humans affect their surroundings and how the environment affects the people who live there. Movement describes how and why people, goods, and ideas move between different places. The region theme examines how places are grouped into different regions. Regions can be divided based on a variety of factors, including physical characteristics, cultures, weather, and political factors.

HOW TO USE THIS BOOK (cont.)

Weekly Themes

The following chart shows the topics, locations, and themes of geography that are covered during each week of instruction.

Week	Topic	Location	Theme(s)
1	—Map Skills Only—		Location
2			Location
3	Landforms	Appalachian Mountains	Place, Region
4	Ecosystem	Mojave Desert	Location, Place, Region
5	Trade	Mississippi River	Human-Environment Interaction, Movement, Region
6	Parks	Stanley Park, Vancouver	Place, Human-Environment Interaction
7	Settlements	San Francisco	Human-Environment Interaction, Movement
8	Landforms	Colorado Plateau	Place, Region
9	Seasons	World	Region
10	Economy	New York	Place, Human-Environment Interaction
11	Gold Rush	North America	Movement
12	Climate	Arctic Tundra	Region
13	Cities	North America	Movement
14	Regions	Great Basin	Place, Region
15	Volcanoes	Hawai'i	Place
16	Government	Harris County, Texas	Location, Place
17	Resources	United States	Human-Environment Interaction, Region

HOW TO USE THIS BOOK *(cont.)*

Week	Topic	Location	Theme(s)
18	Natural Resources	Oregon	Place, Human-Environment Interaction
19	Forced Migration	The Trail of Tears	Movement
20	Culture	Quebec	Location, Place
21	Ecosystem	Florida Everglades	Place, Human-Environment Interaction, Region
22	Biomes	North American Taiga	Location, Place, Region
23	Settlement	United States	Movement, Region
24	Perspectives	Yellowstone National Park	Location, Place
25	Trade	The Caribbean	Location, Place, Region
26	Culture	United States	Location, Place
27	Paul Revere's Ride	Boston	Location, Movement
28	Canyons	Grand Canyon	Place
29	Conservation	New Jersey	Human-Environment Interaction
30	Transportation	North America	Movement
31	Hurrican Katrina	Louisiana	Location, Human-Environment Interaction
32	U.S. Government	Washington, DC	Location, Place
33	Regions	Yukon	Place, Human-Environment interaction, Region
34	Resources	California	Location, Human-Environment Interaction
35	Culture	Mexico City	Location, Place
36	Migration	Florida	Place, Movement

28624—180 Days of Geography

© *Shell Education*

HOW TO USE THIS BOOK *(cont.)*

Using the Practice Pages

The activity pages provide practice and assessment opportunities for each day of the school year. Teachers may wish to prepare packets of weekly practice pages for the classroom or for homework.

As outlined on page 4, each week examines one location and one geography topic.

 The first two days focus on map skills. On Day 1, students will study a map and answer questions about it. On Day 2, they will add to or create a map.

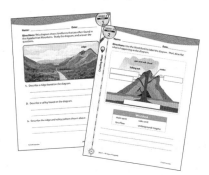

 Days 3 and 4 allow students to apply information and data from texts, charts, graphs, and other sources to the location being studied.

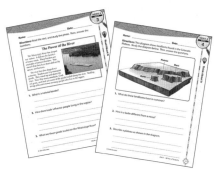

 On Day 5, students will apply what they learned to themselves.

Using the Resources

Rubrics for the types of days (map skills, applying information and data, and making connections) can be found on pages 210–212 and in the Digital Resources. Use the rubrics to assess students' work. Be sure to share these rubrics with students often so that they know what is expected of them.

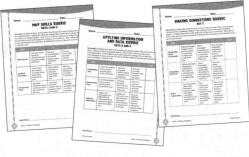

HOW TO USE THIS BOOK *(cont.)*

Diagnostic Assessment

Teachers can use the practice pages as diagnostic assessments. The data analysis tools included with the book enable teachers or parents to quickly score students' work and monitor their progress. Teachers and parents can quickly see which skills students may need to target further to develop proficiency.

Students will learn map skills, how to apply information and data, and how to relate what they learned to themselves. You can assess students' learning in each area using the rubrics on pages 210–212. Then, record their scores on the Practice Page Item Analysis sheets on pages 213–215. These charts are also provided in the Digital Resources as PDFs, Microsoft Word® files, and Microsoft Excel® files (see page 216 for more information). Teachers can input data into the electronic files directly on the computer, or they can print the pages.

To Complete the Practice Page Item Analyses:

- Write or type students' names in the far-left column. Depending on the number of students, more than one copy of the forms may be needed.

    - The skills are indicated across the tops of the pages.

    - The weeks in which students should be assessed are indicated in the first rows of the charts. Students should be assessed at the ends of those weeks.

- Review students' work for the days indicated in the chart. For example, if using the Making Connections Analysis sheet for the first time, review students' work from Day 5 for all five weeks.

- Add the scores for each student. Place that sum in the far right column. Record the class average in the last row. Use these scores as benchmarks to determine how students are performing.

Using the Resources

The Digital Resources contain digital copies of the rubrics, item analysis sheets, and standards charts. See page 216 for more information.

HOW TO USE THIS BOOK *(cont.)*

Using the Results to Differentiate Instruction

Once results are gathered and analyzed, teachers can use them to inform the way they differentiate instruction. The data can help determine which geography skills are the most difficult for students and which students need additional instructional support and continued practice.

Whole-Class Support

The results of the diagnostic analysis may show that the entire class is struggling with certain geography skills. If these concepts have been taught in the past, this indicates that further instruction or reteaching is necessary. If these concepts have not been taught in the past, this data is a great preassessment and may demonstrate that students do not have a working knowledge of the concepts. Thus, careful planning for the length of the unit(s) or lesson(s) must be considered, and additional front-loading may be required.

Small-Group or Individual Support

The results of the diagnostic analysis may show that an individual student or a small group of students is struggling with certain geography skills. If these concepts have been taught in the past, this indicates that further instruction or reteaching is necessary. Consider pulling these students aside to instruct them further on the concepts while others are working independently. Students may also benefit from extra practice using games or computer-based resources.

Teachers can also use the results to help identify proficient individual students or groups of students who are ready for enrichment or above-grade-level instruction. These students may benefit from independent learning contracts or more challenging activities.

STANDARDS CORRELATIONS

Shell Education is committed to producing educational materials that are research and standards based. In this effort, we have correlated all our products to the academic standards of all 50 states, the District of Columbia, the Department of Defense Dependents Schools, and all Canadian provinces.

How to Find Standards Correlations

To print a customized correlation report of this product for your state, visit our website at **www.teachercreatedmaterials.com/administrators/correlations** and follow the on-screen directions. If you require assistance in printing correlation reports, please contact our Customer Service Department at 1-877-777-3450.

Purpose and Intent of Standards

The Every Student Succeeds Act (ESSA) mandates that all states adopt challenging academic standards that help students meet the goal of college and career readiness. While many states already adopted academic standards prior to ESSA, the act continues to hold states accountable for detailed and comprehensive standards. Standards are designed to focus instruction and guide adoption of curricula. Standards are statements that describe the criteria necessary for students to meet specific academic goals. They define the knowledge, skills, and content students should acquire at each level. Standards are also used to develop standardized tests to evaluate students' academic progress. Teachers are required to demonstrate how their lessons meet state standards. State standards are used in the development of our products, so educators can be assured they meet the academic requirements of each state.

The activities in this book are aligned to the National Geography Standards and the McREL standards. The chart on pages 11–12 lists the National Geography Standards used throughout this book. The chart on pages 13–14 correlates the specific McREL and National Geography Standards to each week. The standards charts are also in the Digital Resources (standards.pdf).

C3 Framework

This book also correlates to the College, Career, and Civic Life (C3) Framework published by the National Council for the Social Studies. By completing the activities in this book, students will learn to answer and develop strong questions (Dimension 1), critically think like a geographer (Dimension 2), and effectively choose and use geography resources (Dimension 3). Many activities also encourage students to take informed action within their communities (Dimension 4).

STANDARDS CORRELATIONS *(cont.)*

180 Days of Geography is designed to give students daily practice in geography through engaging activities. Students will learn map skills, how to apply information and data to their understandings of various locations and cultures, and how to apply what they learned to themselves.

Easy to Use and Standards Based

There are 18 National Geography Standards, which fall under six essential elements. Specific expectations are given for fourth grade, eighth grade, and twelfth grade. For this book, fourth grade expectations were used with the understanding that full mastery is not expected until that grade level.

Essential Elements	National Geography Standards
The World in Spatial Terms	**Standard 1:** How to use maps and other geographic representations, geospatial technologies, and spatial thinking to understand and communicate information
	Standard 2: How to use mental maps to organize information about people, places, and environments in a spatial context
	Standard 3: How to analyze the spatial organization of people, places, and environments on Earth's surface
Places and Regions	**Standard 4:** The physical and human characteristics of places
	Standard 5: People create regions to interpret Earth's complexity
	Standard 6: How culture and experience influence people's perceptions of places and regions
Physical Systems	**Standard 7:** The physical processes that shape the patterns of Earth's surface
	Standard 8: The characteristics and spatial distribution of ecosystems and biomes on Earth's surface

STANDARDS CORRELATIONS *(cont.)*

Essential Elements	National Geography Standards
Human Systems	**Standard 9:** The characteristics, distribution, and migration of human populations on Earth's surface
	Standard 10: The characteristics, distribution, and complexity of Earth's cultural mosaics
	Standard 11: The patterns and networks of economic interdependence on Earth's surface
	Standard 12: The process, patterns, and functions of human settlement
	Standard 13: How the forces of cooperation and conflict among people influence the division and control of Earth's surface
Environment and Society	**Standard 14:** How human actions modify the physical environment
	Standard 15: How physical systems affect human systems
	Standard 16: The changes that occur in the meaning, use, distribution, and importance of resources
The Uses of Geography	**Standard 17:** How to apply geography to interpret the past
	Standard 18: How to apply geography to interpret the present and plan for the future

–2012 National Council for Geographic Education

STANDARDS CORRELATIONS *(cont.)*

This chart lists the specific National Geography Standards and McREL standards that are covered each week.

Wk.	NGS	McREL Standards
1	Standard 1	Understands the characteristics and uses of maps, globes, and other geographic tools and technologies.
2	Standards 1 and 2	Understands the characteristics and uses of maps, globes, and other geographic tools and technologies.
3	Standard 4	Knows patterns on the landscape produced by physical processes.
4	Standard 8	Knows plants and animals associated with various vegetation and climatic regions on Earth.
5	Standard 15	Knows how communities benefit from the physical environment.
6	Standard 6	Understands ways in which people view and relate to places and regions differently.
7	Standard 12	Knows the characteristics and locations of cities and how cities have changed over time.
8	Standards 4 and 7	Understands how physical processes help to shape features and patterns on Earth's surface.
9	Standard 7	Knows how Earth's position relative to the sun affects events and conditions on Earth.
10	Standard 14	Knows the ways people alter the physical environment.
11	Standard 9	Knows the relationships between economic activities and resources.
12	Standard 14	Knows plants and animals associated with various vegetation and climatic regions on Earth.
13	Standard 3	Knows the location of major cities in North America.
14	Standard 5	Knows the characteristics of a variety of regions.
15	Standard 7	Understands how physical processes help to shape features and patterns on Earth's surface.
16	Standard 13	Knows the functions of political units and how they differ on the basis of scale.
17	Standard 16	Knows the characteristics, location, and use of renewable resources, flow resources, and nonrenewable resources.
18	Standard 11	Knows economic activities that use natural resources in the local region, state, and nation and the importance of the activities to these areas.

STANDARDS CORRELATIONS *(cont.)*

Wk.	NGS	McREL Standards
19	Standard 9	Understands voluntary and involuntary migration. Knows the causes and effects of human migration.
20	Standard 10	Knows the similarities and differences in characteristics of culture in different regions.
21	Standard 14	Knows the ways people alter the physical environment.
22	Standard 8	Knows plants and animals associated with various vegetation and climatic regions on Earth.
23	Standard 17	Knows the factors that have contributed to changing land use in a community.
24	Standard 6	Understands ways in which people view and relate to places and regions differently.
25	Standard 11	Knows how regions are linked economically and how trade affects the way people earn their living in each region.
26	Standard 10	Understands cultural change.
27	Standard 17	Knows the geographic factors that have influenced people and events in the past.
28	Standard 7	Understands how physical processes help to shape features and patterns on Earth's surface.
29	Standard 18	Knows human-induced changes that are taking place in different regions and the possible future impacts of these changes.
30	Standard 11	Knows the major transportation routes that link resources with consumers and the transportation modes used.
31	Standard 15	Knows natural hazards that occur in the physical environment.
32	Standard 13	Knows the functions of political units and how they differ on the basis of scale.
33	Standard 15	Knows how communities benefit from the physical environment. Knows the ways in which human activities are constrained by the physical environment.
34	Standard 16	Knows the relationship between population growth and resource use. Knows the ways in which resources can be managed and why it is important to do so.
35	Standard 10	Knows the similarities and differences in characteristics of culture in different regions.
36	Standard 9	Knows the causes and effects of human migration.

Name: _____ **Date:** _____

Directions: A compass rose tells you directions on a map. Use the compass rose to answer the questions about the Hawai'ian Islands.

The Hawai'ian Islands

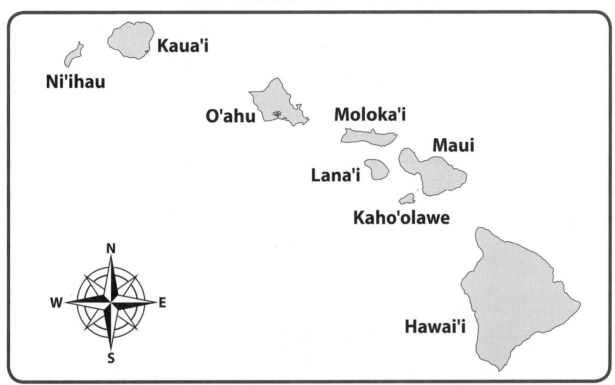

1. Name the closest island north of Lana'i.

   _____

2. Name the closest island northwest of Moloka'i.

   _____

3. Name the island that is southeast of Maui.

   _____

4. Name the closest island northwest of O'ahu.

   _____

Map Skills

Name: _____ **Date:** _____

Directions: Study the map of Tennessee. Then, answer the questions.

Tennessee

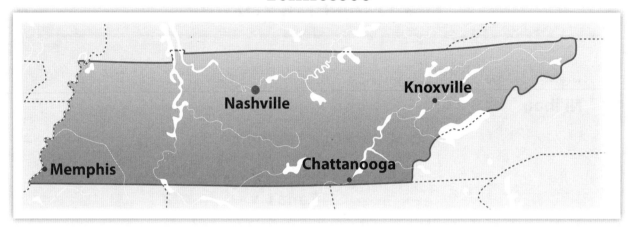

1. Label the compass rose with the four cardinal directions and the four intermediate directions.

2. Chattanooga is southeast of which city in Tennessee?

   _____

3. Knoxville is northeast of which city in Tennessee?

   _____

4. Which direction would you go to travel from Chattanooga to Memphis?

   _____

Name: _____ Date:_____

Directions: This map uses shading to show population. Study the map and legend closely. Then, answer the questions.

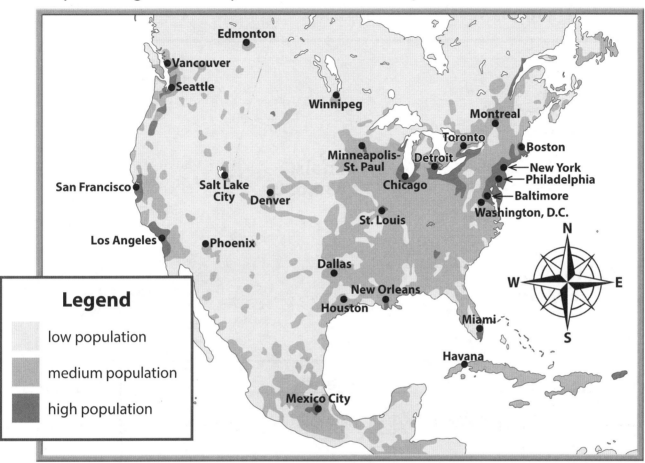

Map Skills

Legend

low population

medium population

high population

1. Circle the part of the legend that describes areas with the lowest population.

2. How does the legend explain what the shades mean?

   _____

   _____

3. Why might major cities be shaded the darkest?

   _____

   _____

Map Skills

Name: _____ Date: _____

Directions: This map shows the five states with the most people. Color each of the five states a different color. Color the legend to match. Then, answer the questions.

N
W E
S

New York:
19,378,102

Illinois:
12,830,632

California:
37,253,956

Texas:
25,145,561

Florida:
18,801,310

Legend

☐ California   ☐ New York   ☐ Florida

☐ Illinois   ☐ Texas

1. Which state has the highest population?

_____

2. Which of the five states is found in the Northeast?

_____

3. List the five states from east to west.

_____

_____

Name: _____ **Date:**_____

Directions: Create a map of your classroom. Your map should have a title, legend, and compass rose.

Map Skills

Map Skills

Name: _____   Date:_____

Directions: Study the globe. Then, answer the questions.

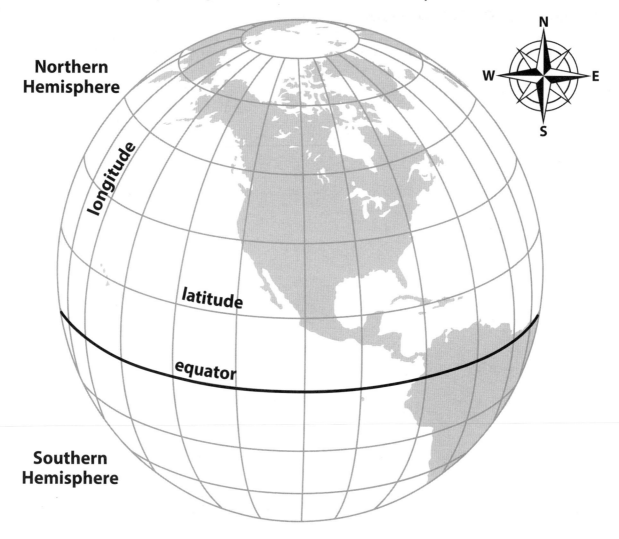

Northern
Hemisphere

longitude

latitude

equator

Southern
Hemisphere

N
W   E
S

1. What separates the Northern Hemisphere from the Southern
    Hemisphere?

    _____

2. Which lines run from north to south?

    _____

3. Which lines run from west to east?

    _____

Name: _____ **Date:** _____

Directions: Study the globe. Then, follow the steps.

1. Label a line of latitude.

2. Label a line of longitude.

3. Label the equator.

4. Trace all the lines of latitude red.

5. Trace all the lines of longitude blue.

Map Skills

Name: _____ **Date:** _____

Directions: This is a community map. The grid helps people find different places. Use the map to answer the questions.

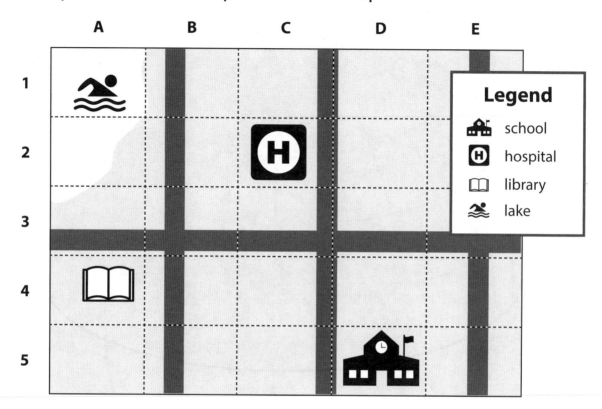

1. What is located in A1?

   _____

2. In which square is the school located?

   _____

3. In which square is the hospital located?

   _____

4. What is located in A4?

   _____

Name: _____ **Date:** _____

Directions: Map legends often include symbols. These symbols help people read maps. Study the map of Georgia. Then, answer the questions.

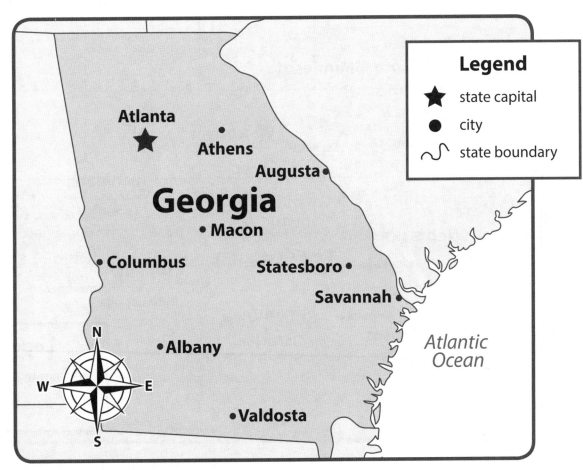

1. What is the capital of Georgia?

_____

2. List at least three cities found in Georgia.

_____

_____

3. Which city is found on the western border of Georgia?

_____

Map Skills

Name: _____ Date:_____

Directions: Use the grid and the legend to answer the questions.

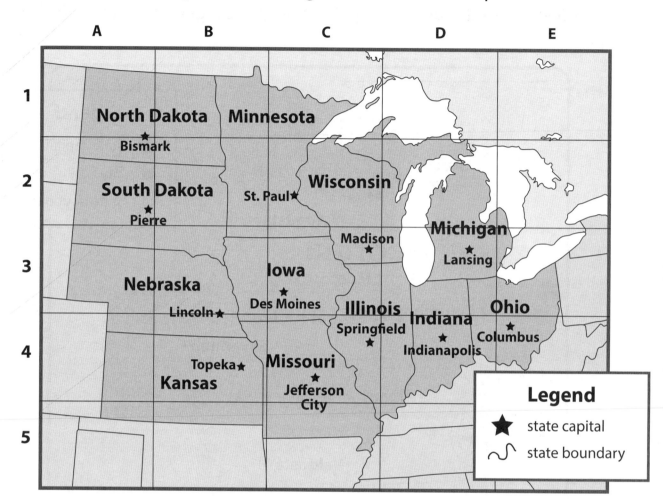

1. What is the capital of Kansas?

   _____

2. In which square is Madison, Wisconsin, located?

   _____

3. In which square is Pierre, South Dakota, located?

   _____

Name: _____ **Date:** _____

Directions: This diagram shows landforms that are often found in the Appalachian Mountains. Study the diagram, and answer the questions.

1. Describe a ridge based on the diagram.

_____

_____

2. Describe a valley based on the diagram.

_____

_____

3. Describe the ridge and valley pattern shown above.

_____

_____

_____

Reading Maps

Name: _____ **Date:**_____

Directions: Study the photo closely. Then, create a diagram based on the photo. Label the ridges and valleys in your diagram.

Creating Maps

Name: _____ **Date:**_____

Directions: Read the text, and study the photo. Then, answer the questions.

Shaping Ridges and Valleys

Landforms are shaped by many forces. One force is water. Water can wear down land over time. This is called *erosion*. Ridges and valleys are a great example of water's power.

Some rocks are harder to wear down than others. Ridges are made of sturdier rock, such as sandstone. This rock is hard to wear down. Valleys are made of softer rock, such as limestone. Limestone can be worn away by water more quickly.

1. What is one force that helps shape ridges and valleys?

   _____

2. How is sandstone different from limestone?

   _____

   _____

   _____

3. How might erosion have shaped the ridges and valleys in the photo?

   _____

   _____

   _____

Name: _____ **Date:**_____

Directions: Study the map, and answer the questions.

Think About It

Appalachian Mountain Regions

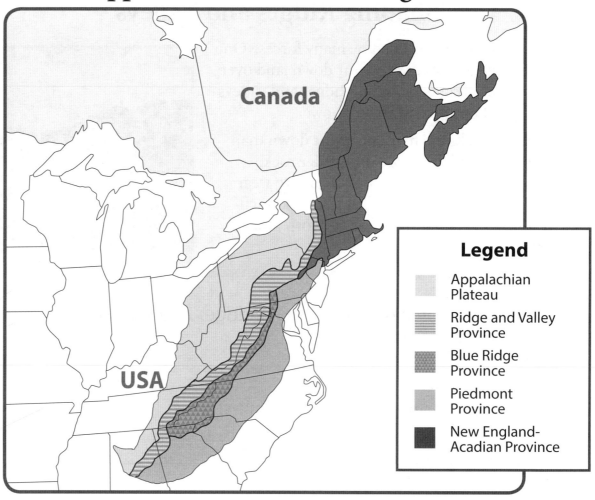

Canada

USA

Legend

- Appalachian Plateau
- Ridge and Valley Province
- Blue Ridge Province
- Piedmont Province
- New England-Acadian Province

1. The Blue Ridge Province is between which two provinces?

_____

_____

2. Outline the states that include at least four provinces.

3. Which province extends into Canada?

_____

Name: _____ **Date:**_____

Directions: Write a type of landform that is near you. Complete the Venn diagram to compare and contrast it with Appalachian valleys.

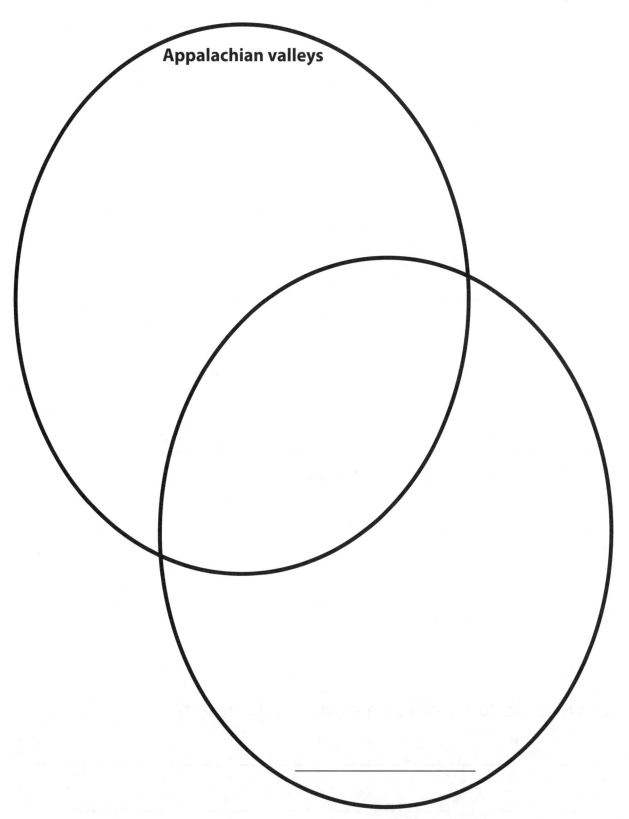

Appalachian valleys

_____

Reading Maps

Name: _____   Date:_____

Directions: Study the map closely. Then, answer the questions.

Mojave Desert

1. The Mojave Desert is in which states?

    _____

    _____

2. In which state is most of the desert found?

    _____

3. Which California cities are located in the desert?

    _____

    _____

Name: _____ **Date:** _____

Directions: Label the states that contain the Mojave Desert. Then, use the legend to answer the questions.

Mojave Desert

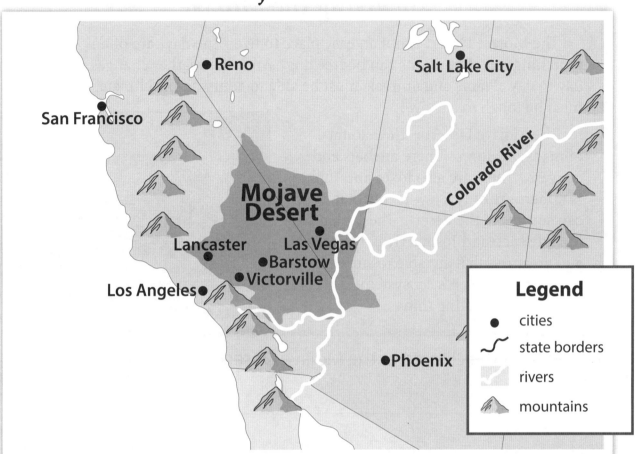

1. What river is labeled on the map?

_____

2. Are there mountains in California? How do you know?

_____

_____

3. What city is labeled in Arizona?

_____

Name: _____ **Date:** _____

Read About It

Directions: Read the text, and study the photo. Then, answer the questions.

Surviving the Mojave

The Mojave Desert is not an easy place to live. The days are often scorching hot. But nights can be freezing. And the weather changes from day to day. Plants and animals must be able to adjust to the climate. They must be tough to survive!

The Mojave Desert is the habitat of many creatures. There are bats and cougars. There are also foxes and snakes. But there are only a few plants that can survive in the desert. These plants include cacti and Joshua trees. The plants and creatures that are able to survive make up the desert's ecosystem. This is a community of living things.

1. Describe the climate of the Mojave Desert.

_____

_____

2. Why can only a few plants survive in the desert?

_____

_____

3. What is an ecosystem?

_____

_____

Name: _____ **Date:** _____

Directions: This chart shows the monthly high and low temperatures in Barstow, California. This city is in the Mojave Desert. Study the chart, and answer the questions.

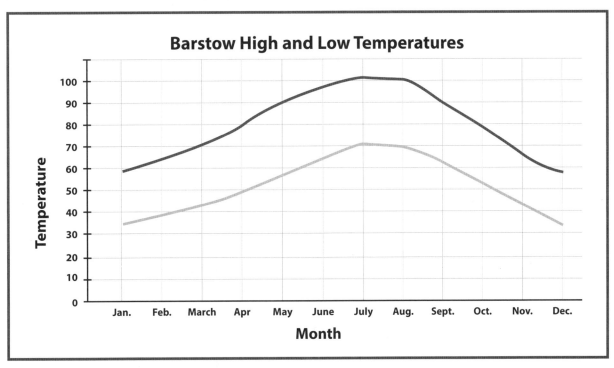

1. Which month had the highest temperatures?

   _____

2. Which two months had the lowest temperatures?

   _____

3. If you were going to visit the Mojave Desert, in what month would you go? Why?

   _____

   _____

   _____

Think About It

Name: _____ **Date:** _____

Geography and Me

Directions: Create a map of your local ecosystem. Include plants and animals that are found in your community.

Name: _____   **Date:**_____

Directions: The Mississippi River is the longest river in North America. Closely study the map. Then, answer the questions.

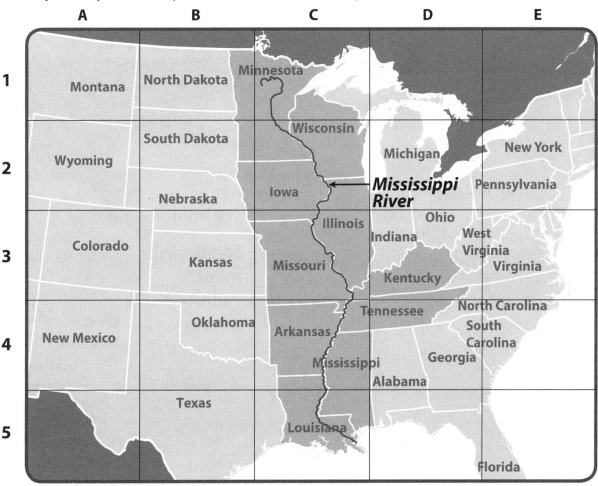

1. In which square does the Mississippi River begin?

_____

2. In which square does the Mississippi River form Arkansas's border?

_____

3. List at least five states that border the river.

_____

_____

Creating Maps

Name: _____ **Date:**_____

Directions: Follow the steps to complete the map.

_____

1. Trace the Mississippi River in blue. Circle where it starts and ends.

2. Add a title and compass rose to the map.

Name: _____ **Date:** _____

Directions: Read the text, and study the photo. Then, answer the questions.

The Power of the River

The Mississippi River has shaped our nation. It defines the borders of many states. These borders are examples of natural borders. This means that they have been defined by nature, not humans.

The river has also shaped our economy. Traders use it to transport goods up and down the river. They ship many products. Coal, oil, grain, and steel are some of the goods traded along the river. Trading brings jobs and money to many people who live in the region.

1. What is a natural border?

_____

_____

2. How does trade influence people living in the region?

_____

_____

_____

3. What are three goods traded on the Mississippi River?

_____

_____

Name: _____ **Date:** _____

Think About It

Directions: This map shows the river ports in Mississippi. Ports are places where ships load and unload goods. They are important to the economy of the surrounding cities.

River Ports

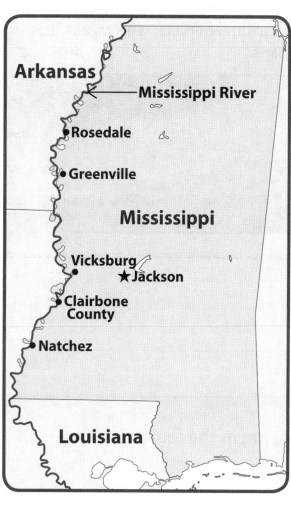

1. Which two ports can be found on the Arkansas and Mississippi state border?

   _____

   _____

2. How does the Mississippi River relate to Mississippi's border?

   _____

   _____

   _____

   _____

   _____

3. If you were shipping goods from Jackson, which port would you use? Why?

   _____

   _____

   _____

Name: _____ **Date:** _____

Directions: Imagine you are a trader on the Mississippi River. You are traveling from Missouri to Wisconsin. Draw the route you will take along the river. Then, write a paragraph telling about your journey.

_____

_____

_____

_____

_____

_____

Reading Maps

Name: _____ **Date:** _____

Directions: Study this map of a large park. Then, answer the questions.

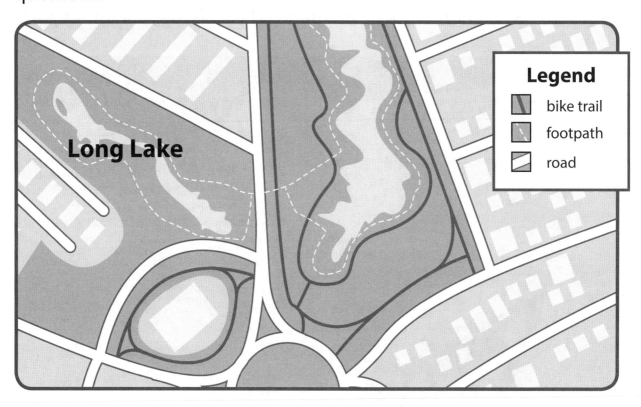

Long Lake

Legend
- bike trail
- footpath
- road

1. What are three modes of transportation you can use to explore the park?

   _____

   _____

2. If you wanted to take your time exploring a trail, which mode of transportation would you use?

   _____

3. If you wanted to explore Long Lake, which form of transportation would you use?

   _____

Name: _____ **Date:**_____

Directions: Follow the steps, and answer the questions.

Long Lake

Legend

	bike trail
	footpath
	road

1. Trace the route you would take to explore the park by bike.

2. Trace the route you would take to explore the park by foot.

3. Compare your two routes. How would your experience biking through the park be different from walking through the park?

_____

_____

4. How might your experience be different if you drove through the park?

_____

_____

Read About It

Name: _____ **Date:** _____

Directions: Read the text, and study the photo. Then, answer the questions.

Fun for Everyone

Stanley Park is a landmark in Vancouver, Canada. It is a large park with trails, lakes, and roads. There is much to do at Stanley Park. Everyone can find a park activity they love. Some use the park to exercise. They jog on trails and ride along bike paths. Others use the park to learn. They take tours to learn about the park's plants and animals.

The park is also a place for athletes! There are tennis courts and golf courses. People can play their favorite sports while surrounded by nature. Some people sail boats around the park. Stanley Park has something for everyone!

1. What might a student think about Stanley Park?

_____

_____

2. What might an athlete think about Stanley Park?

_____

_____

3. Why might people have different views about the same park?

_____

_____

Name: _____ **Date:**_____

Directions: The leaders of Vancouver want to plant 150,000 new trees by 2020. Study the graph. Then, answer the questions.

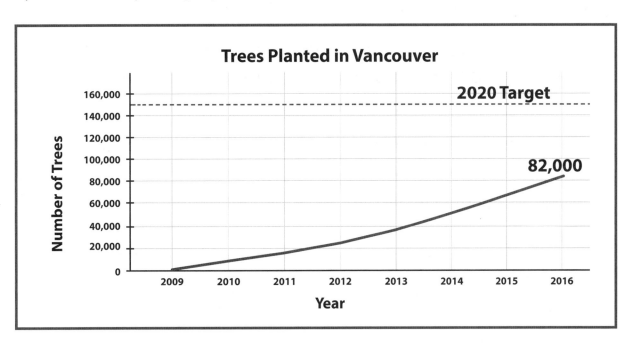

Trees Planted in Vancouver

Think About It

1. About how many new trees had been planted by 2012?

   _____

2. About how many trees were planted between 2012 and 2014?

   _____

3. In 2016, how many more trees still needed to be planted to meet the 2020 goal?

   _____

4. Do you think Vancouver leaders will meet their goal by 2020? Why or why not?

   _____

   _____

Geography and Me

Name: _____ **Date:** _____

Directions: Imagine you are spending a day at Stanley Park. You have decided to create a schedule. Use the Word Bank to help you complete each sentence below.

Word Bank				
explore	bike paths	tour	play	picnic
tennis courts	hiking trails	plants	learn	animals

When I first get to the park, I will

_____

_____ .

Then, I will

_____

_____ .

Next, I will take a break, and

_____

_____ .

Finally, I will

_____

_____ .

Name: _____ **Date:** _____

Directions: Study the map, and answer the questions.

San Francisco Landmarks

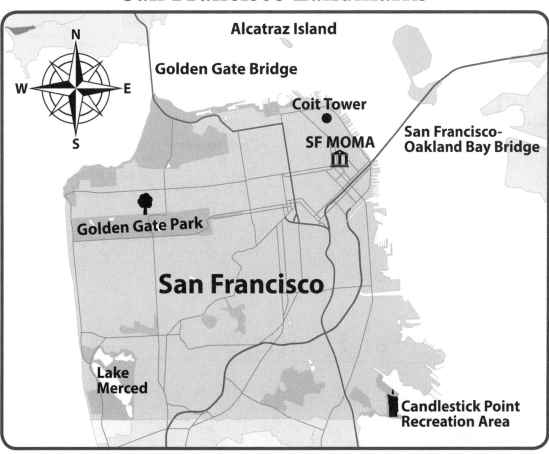

1. What two bridges can you take to reach San Francisco?

_____

_____

2. Golden Gate Park is south of which bridge?

_____

3. Name two attractions in San Francisco.

_____

Creating Maps

Name: _____ **Date:** _____

Directions: This map has symbols for different city landmarks. Create a legend with symbols to help readers understand the map.

San Francisco Landmarks

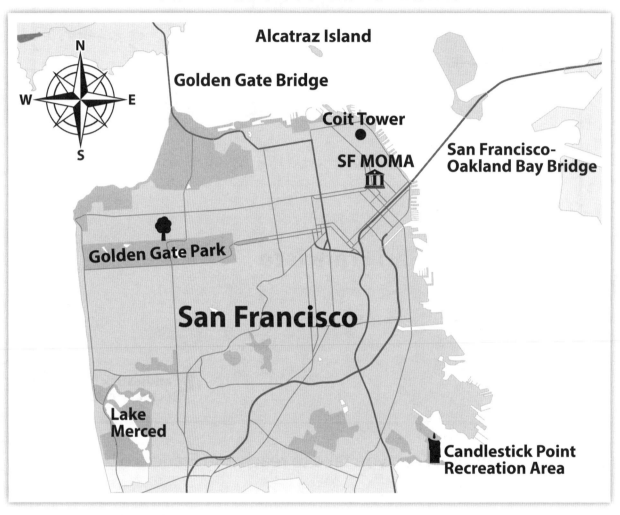

```
Legend
```

Name: _____ **Date:** _____

Directions: Read the text, and study the photo. Then, answer the questions.

Living in the City

When you live in a city, the options are endless. Day or night, there are things to do. The city is always awake, filled with noise and light. People move in crowds, eager to find the next fun activity.

When you're hungry, there are hundreds of restaurants to choose from. When you want to see a movie, there is likely a theater around the corner. If art is your thing, just head to a museum. Living in the city is convenient. Stores open early and close late. No matter what you need, there is likely a place nearby that provides it!

1. Name three things you can do in a city.

   _____

   _____

2. What are some benefits of living in a city?

   _____

   _____

3. What might be some drawbacks to living in the city?

   _____

   _____

Think About It

Name: _____ Date: _____

Directions: Study the map. Then, answer the questions.

San Francisco Subways

1. How many subway lines pass through San Francisco?

   _____

2. How many subway lines stop at MacArthur?

   _____

3. Why might so many subway lines pass through San Francisco?

   _____

   _____

Name: _____ **Date:** _____

Directions: Complete the Venn diagram to compare and contrast your community with San Francisco.

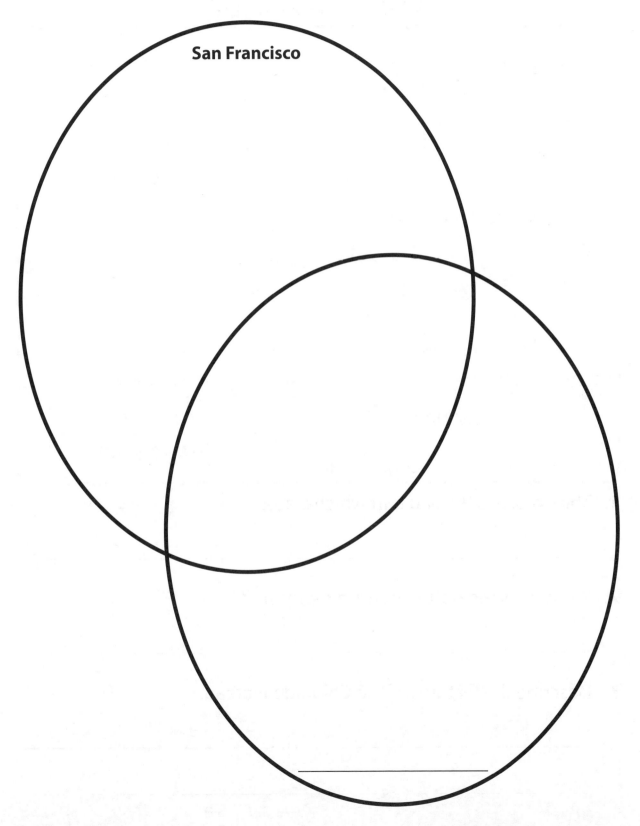

San Francisco

_____

Reading Maps

Name: _____ **Date:** _____

Directions: Study the map closely. Then, answer the questions.

Colorado Plateau

1. The Colorado Plateau is in which states?

_____

2. What does the dotted pattern represent?

_____

3. Describe the features of the Colorado Plateau.

_____

_____

© Shell Education

Name: _____ **Date:** _____

Directions: Follow the steps to complete the map.

_____

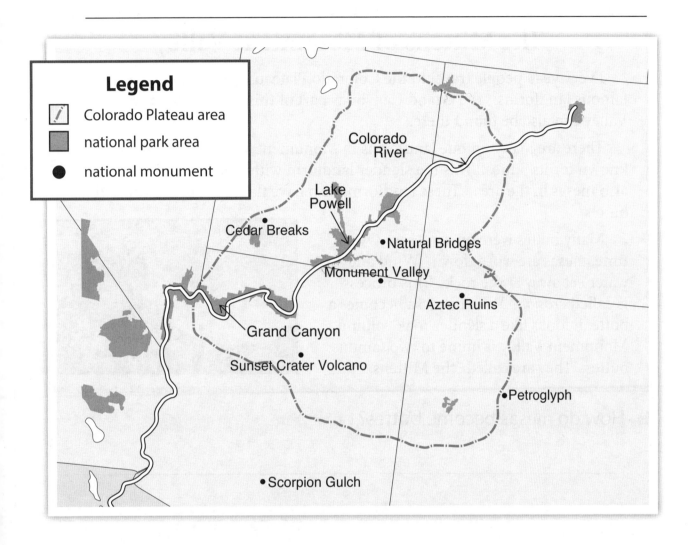

Legend

/ Colorado Plateau area

national park area

● national monument

Colorado River

Lake Powell

Cedar Breaks

Natural Bridges

Monument Valley

Aztec Ruins

Grand Canyon

Sunset Crater Volcano

Petroglyph

Scorpion Gulch

1. Shade the Colorado Plateau.

2. Trace the Colorado River in blue.

3. Label the states in the Colorado Plateau.

4. Add a title to the map.

5. Add a compass rose to the map.

Creating Maps

Name: _____ Date:_____

Read About It

Directions: Read the text, and study the photo. Then, answer the questions.

The Mittens of Monument Valley

Every year, people travel to the Colorado Plateau. They want to see its famous landforms. The Grand Canyon is part of this region. Monument Valley can also be found there.

There are many unique landforms in Monument Valley. The area is known for its buttes. This is a slender landform with a flat top. There are also mesas in the area. These landforms also have flat tops but are wider than buttes.

Many buttes were once mesas. Over time, mesas are worn down. Wind and water eat away at the rock. This process is called *erosion*. When a mesa becomes a butte, it looks like a slender rock column. Monument Valley is home to two famous buttes. They are called "the Mittens."

1. How do mesas become buttes?

_____

_____

2. What is erosion?

_____

_____

3. Why do you think these landforms are called "mittens"?

_____

_____

Name: _____ **Date:** _____

Directions: This diagram shows landforms found in the Colorado Plateau. Study the diagram. Then, answer the questions.

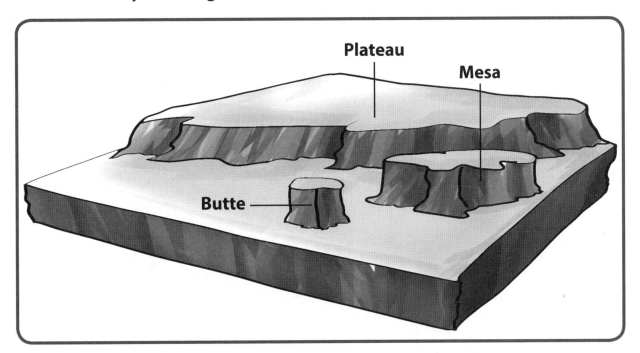

1. What do these landforms have in common?

   _____

   _____

2. How is a butte different from a mesa?

   _____

   _____

3. Describe a plateau as shown in the diagram.

   _____

   _____

   _____

Think About It

Name: _____ **Date:** _____

Directions: Imagine you are visiting the Colorado Plateau. Draw the landforms you might see there.

Geography and Me

28624—180 Days of Geography

© Shell Education

Name: _____   **Date:** _____

Directions: This diagram shows the tilt of Earth during December and June. Study the diagram, and answer the questions.

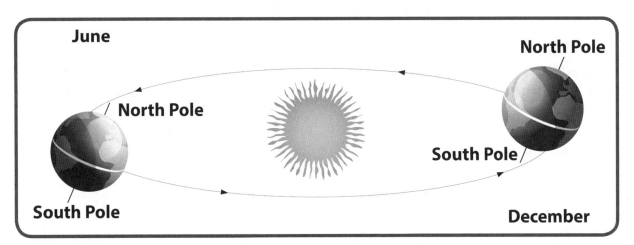

1. Does the Northern Hemisphere get more sunlight in June or December? How do you know?

   _____

   _____

2. Does the Southern Hemisphere get more sunlight in June or December? How do you know?

   _____

   _____

3. Do you think December is a warm month or a cold month in the Southern Hemisphere? Why?

   _____

   _____

   _____

Creating Maps

Name: _____ **Date:** _____

Directions: This diagram shows the Northern Hemisphere in all four seasons. Label the four seasons of the Southern Hemisphere. They are opposite from the Northern Hemisphere.

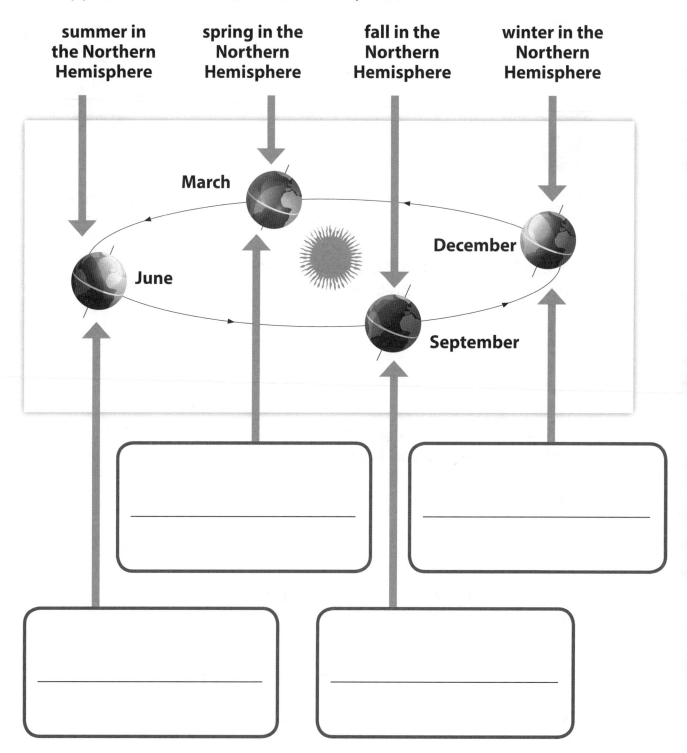

Name: _____ **Date:** _____

Directions: Read the text, and study the diagram. Then, answer the questions.

The Power of Earth's Tilt

Each year, Earth revolves around the sun. As months pass by, different parts of Earth receive direct sunlight. This is what creates our seasons.

For half of the year, the Northern Hemisphere is tilted toward the sun. During this time of year, it receives direct sunlight. These are warm months. They are the seasons of spring and summer. For the other half of the year, the North is tilted away from the sun. These are the cold months of fall and winter.

The Southern Hemisphere is the opposite! When the North is tilted away from the sun, the South is tilted toward the sun. That is why when it is winter north of the equator, it is summer south of the equator.

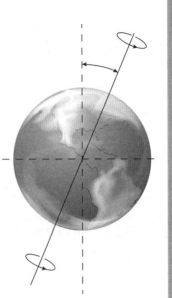

1. What do you think the word *revolves* means?

   _____

   _____

2. What happens when the Northern Hemisphere is tilted away from the sun?

   _____

   _____

3. How are the Northern and Southern Hemispheres opposites?

   _____

   _____

Think About It

Name: _____   Date:_____

Directions: This chart shows the average temperatures of two cities in December and June. Study the chart closely. Then, answer the questions.

City Name	Average High in December	Average Low in December	Average High in June	Average Low in June
Puerto Natales	63°F (17°C)	43°F (6°C)	41°F (5°C)	28°F (-2°C)
Inuvik	-6°F (-21°C)	-23°F (-30°C)	60°F (15°C)	40°F (4°C)

1. Based on the chart, do you think Puerto Natales is in the Southern Hemisphere or Northern Hemisphere? Why?

   _____

   _____

2. Based on the chart, do you think Inuvik is in the Southern Hemisphere or Northern Hemisphere? Why?

   _____

   _____

3. Will the temperature in Puerto Natales likely increase or decrease in July? Why?

   _____

   _____

   _____

Name: _____ **Date:** _____

Directions: Circle where you live on the map. Then, use the Word Bank to complete the sentences.

Equator

Word Bank			
Southern	Northern	cold	hot
warm	cool	tilt	revolve

1. I live in the _____ Hemisphere.

2. In June, the weather is _____.

3. In December, the weather is _____.

4. This is because of the _____ of the Earth.

Reading Maps

Name: _____ **Date:**_____

Directions: Study the map, and answer the questions.

New York Industries

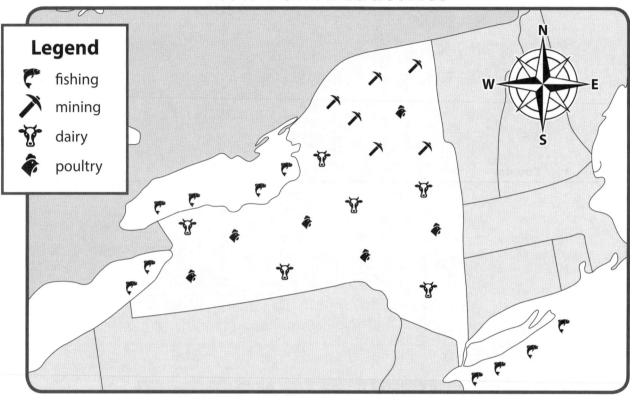

Legend
- 🐟 fishing
- ⛏ mining
- 🐄 dairy
- 🐔 poultry

1. Which industry takes place mostly in the northern part of the state?

   _____

2. Which industry is busiest along the coasts of the Great Lakes and the Atlantic Ocean?

   _____

3. Which industries are found throughout the state?

   _____

Name: _____ **Date:** _____

Directions: Follow the steps to complete the map.

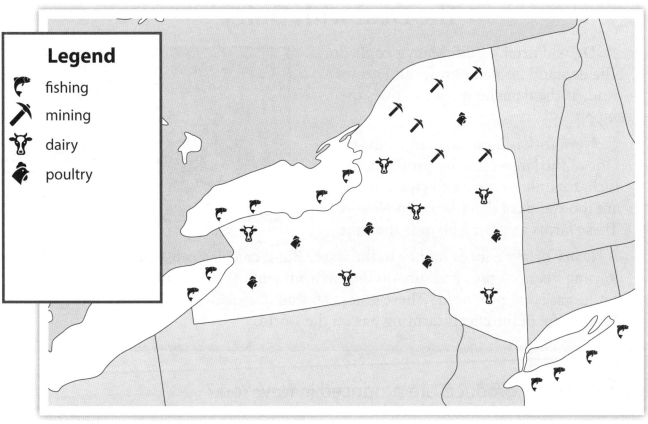

Legend
- 🐟 fishing
- ⛏ mining
- 🐄 dairy
- 🐔 poultry

1. Add a compass rose to the map.

2 Wheat and corn are also grown in the eastern part of the state. Create symbols for these goods, and add them to the legend.

3. Add your symbols to the eastern part of the map.

4. Add a title to the map.

Challenge: Label any states, bodies of water, or countries on the map that you know.

Read About It

Name: _____ **Date:** _____

Directions: Read the text, and study the photo. Then, answer the questions.

The Deal with Dairy

Do you drink milk? Many people do. The demand for dairy grows and grows. And, as the demand grows, so does the supply.

New York is home to many industries. One of the largest is dairy products. This includes milk, yogurt, and cheese. There are thousands of dairy farms in New York. These farms are found all over the state.

Dairy brings a lot of money to the state. But it can also cause problems. Raising livestock puts pressure on the environment. When farmers raise cattle, gases are produced. These gases can lead to global warming. We must be mindful of the effects farming has on the world.

1. What dairy products are produced in New York?

_____

_____

2. Why are there so many dairy farms?

_____

_____

3. How does raising livestock affect the environment?

_____

_____

Name: _____ **Date:** _____

Directions: This chart shows different products farmed in New York. Study the chart, and answer the questions.

Product	Harvested Land (acres)	Total Yield	Total Value
soybeans	320,000	41 bushels per acre	$125,296,000
wheat	115,000	74 bushels per acre	$34,466,000
oats	60,000	55 bushels per acre	$7,425,000

Think About It

1. Which product covered the most farmland?

_____

2. Which product yielded the most bushels per acre?

_____

3. Which product made the most money?

_____

4. If you were starting a new farm in New York, which of these crops would you rather grow? Why?

_____

_____

_____

_____

Geography and Me

Name: _____ **Date:** _____

Directions: Think about the things that are bought and sold in your area. Draw a map of local industries. Remember to include a legend!

Name: _____ **Date:** _____

Directions: The California Gold Rush started in 1848. This map shows the main routes people took to reach California. Study the map, and answer the questions.

1. What troubles do you think gold miners faced on the overland route?

_____

_____

_____

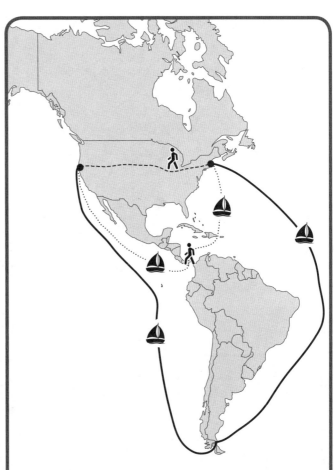

2. What troubles do you think gold miners faced on the Cape Horn route?

_____

_____

_____

3. Which of the three routes would you have taken? Why?

_____

_____

_____

_____

Legend

 overland route— four to six months

 Panama route— one to three months

 Cape Horn route— six to eight months

Creating Maps

Name: _____ **Date:** _____

Directions: Use three different colors or patterns to draw the three routes gold miners took to reach California. Add the colors or patterns you used to the legend.

Legend

[ ] overland route—
four to six months

[ ] Panama route—
one month

[ ] Cape Horn route—
six to eight months

28624—180 Days of Geography   © Shell Education

Name: _____ **Date:** _____

Directions: Read the text, and study the photo. Then, answer the questions.

Gold Fever

Our story begins at Sutter's Mill. It was there, in 1848, that gold was first found in California. Many people didn't believe that the gold was real. In the past, there had been talk about gold, but it had not been true. But when more people found gold, the country caught gold fever!

Gold miners traveled from all over. They were eager to try their luck. Some traveled thousands of miles. In just a few years, hundreds of thousands of people had migrated. During that time, miners found millions of dollars in gold.

But life in the mining camps was rough. There was no one to govern the miners. Life could often be violent. Mining was hard work, too. Miners used simple tools to hunt for gold. They used picks to break away rock and dirt. Then, miners filled pans with dirt and water. They sifted the dirt from the pan, hoping to find pieces of gold. Miners also used tools called *cradles*, which were able to sift through dirt faster.

1. What happened when gold was first discovered at Sutter's Mill?

   _____

   _____

2. What struggles did gold miners face?

   _____

   _____

3. What tools did miners use?

   _____

Think About It

Name: _____ **Date:** _____

Directions: This is a diagram of a mining tool called a *cradle*. It could sift through dirt faster than panning. Study the diagram. Then, answer the questions.

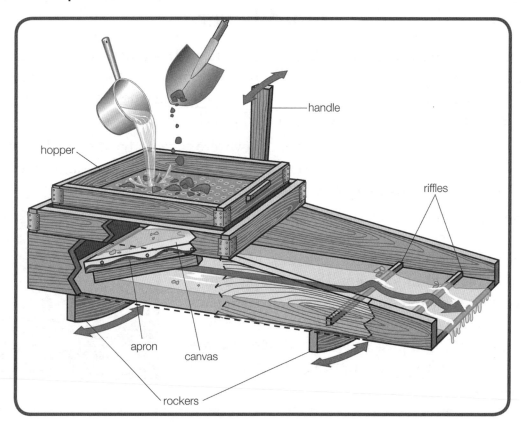

1. Where would the gold collect?

   _____

2. Why might this tool work faster than panning?

   _____

   _____

3. Why might miners want a faster tool?

   _____

   _____

Name: _____ **Date:** _____

Directions: Imagine you are a gold miner traveling to California in 1849. Plot your route on the map. Then, write about your experiences panning for gold.

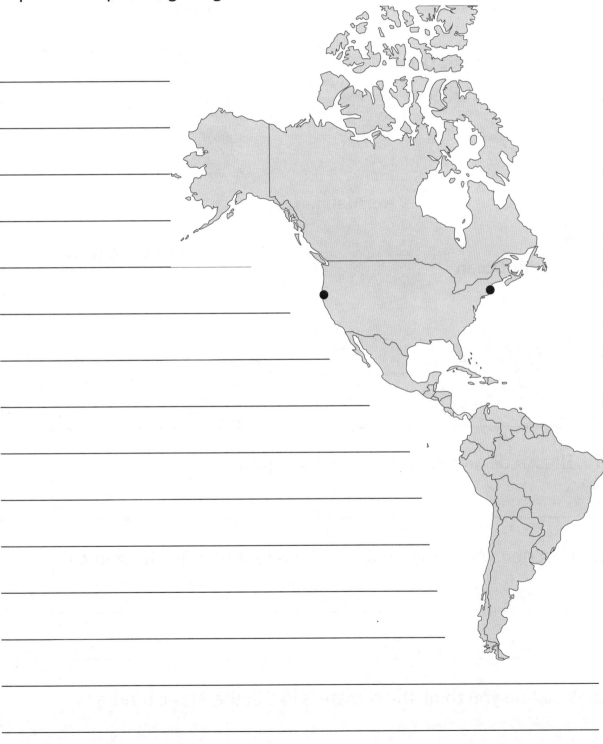

_____

_____

_____

_____

_____

_____

_____

_____

_____

_____

_____

_____

_____

_____

_____

Reading Maps

Name: _____ **Date:** _____

Directions: Study the map. Then, answer the questions.

Arctic Tundra in North America

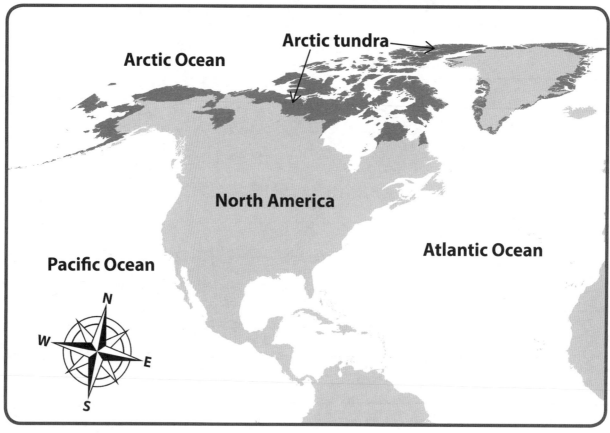

1. The Arctic tundra is close to which ocean?

   _____

2. Describe where the Arctic tundra is found in North America.

   _____

   _____

3. What do you think the climate is like in the Arctic tundra?

   _____

   _____

Name: _____ **Date:**_____

Directions: Follow the steps to complete the map.

Arctic Tundra in North America

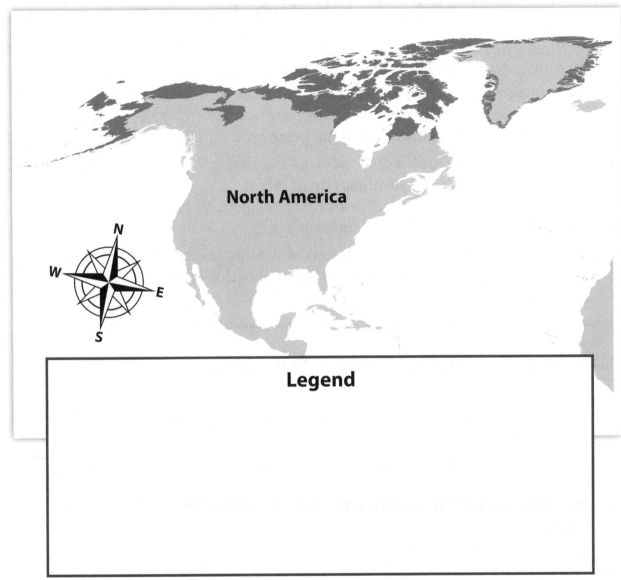

1. Snowy owls and polar bears live in the Arctic tundra. Create a symbol for each animal, and add them to the map.

2. Create a legend to show what your symbols represent.

3. Trace the Arctic tundra region in blue. Add the blue color to your legend.

Name: _____ Date:_____

Directions: Read the text, and study the photo. Then, answer the questions.

Read About It

Living on Thin Ice

The Arctic tundra is a treeless, icy land. During the winter, the temperature can fall to -25°F (-32°C). The summers are warmer, but still cold. The land is home to creatures such as the snowy owl, Arctic fox, and reindeer. It is also the home of the polar bear.

Polar bears are made for their habitat. They have thick fur and a layer of fat that helps them retain heat. Uneven skin on their feet helps them balance on slippery ice. Sharp claws help them snatch prey.

Polar bears live on ice floes, or sheets of floating sea ice. The bears need these ice floes to survive. Sadly, scientists predict that most ice floes will disappear in the future. People are working to help save polar bears.

1. How has the polar bear adapted to its habitat?

_____

_____

2. In what parts of the world would you not find the polar bear? Why?

_____

_____

3. How might a polar bear's white fur help it survive?

_____

_____

Name: _____ **Date:** _____

Directions: This chart shows the amount of Arctic sea ice. Study the chart, and answer the questions.

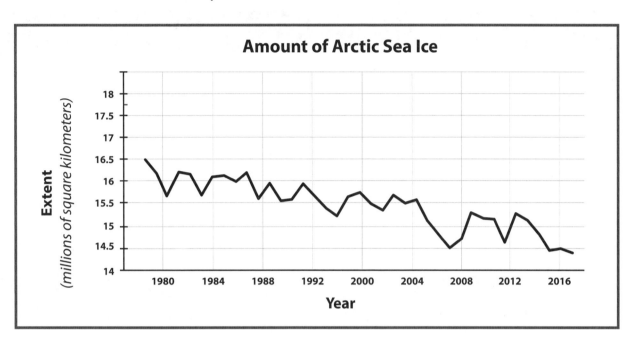

Amount of Arctic Sea Ice

Extent
(millions of square kilometers)

Year

1. Which year shown had the most sea ice?

    _____

2. What trend does this chart show about sea ice?

    _____

    _____

3. How might this sea ice trend affect the polar bear?

    _____

    _____

    _____

Geography and Me

Name: _____ **Date:**_____

Directions: Compare where you live to the Arctic tundra. Write about the climate in each location.

My Home	Arctic Tundra

Name: _____ **Date:** _____

Directions: Study the map closely. Then, answer the questions.

Population of North America

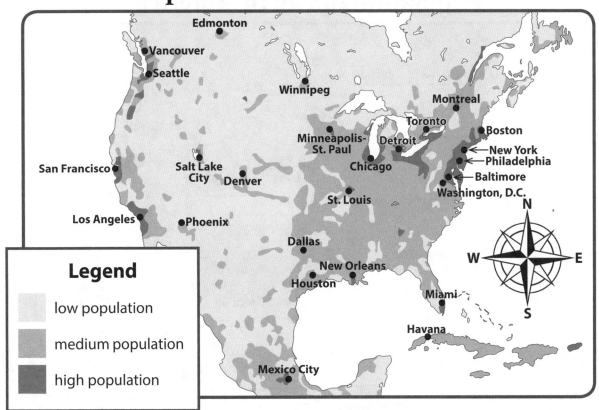

Legend

low population

medium population

high population

1. How does Canada's population compare to the United States?

   _____

   _____

2. List three cities that have high big populations.

   _____

3. Do more people prefer to live on the coast or inland? How do you know?

   _____

   _____

Creating Maps

Name: _____ Date:_____

Directions: Use the clues to label the missing cities.

Population of North America

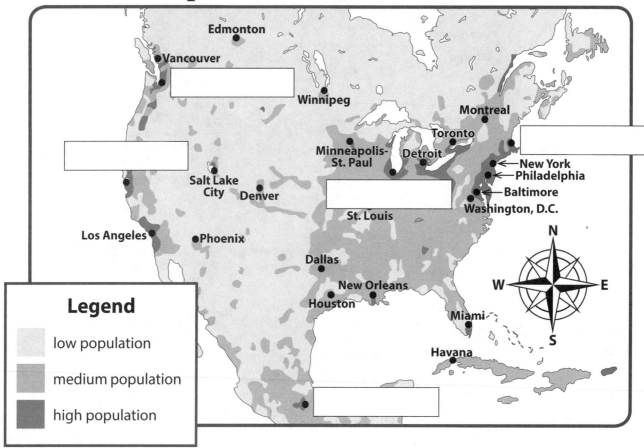

1. Boston is on the East Coast.

2. San Francisco is on the West Coast.

3. Seattle is north of San Francisco.

4. Chicago is in the Midwest.

5. Mexico City is the farthest south.

Challenge: Circle places on the map that you think are big cities. How many places did you circle?

Name: _____ **Date:**_____

Directions: Read the text, and study the photo. Then, answer the questions.

Life on the Coasts

When you live near the water, life can be lovely. There is much to see and do. People swim, surf, and play water sports. Nature is diverse. There is much to explore in the world of plants and animals. These are some of the reasons people choose to live on the coast.

But it's not just nature that draws people to move. History shows us that ports are centers of jobs and trade. Cities grow out from these ports. For example, both Boston and New York City began as ports! Over time, these communities continued to grow. As they grew, their cultures became more dynamic and diverse. New businesses were started. Restaurants, theaters, museums, and concert halls were built. These attractions brought even more people! Our greatest cities continue to grow every day.

1. How are ports related to cities?

_____

_____

2. What happens when cities grow?

_____

_____

3. Why are people drawn to the coasts?

_____

_____

Think About It

Name: _____ **Date:** _____

Directions: This chart shows how the populations of five U.S. cities have changed. Study the chart, and answer the questions.

City	Population in 2000	Population in 2010	Change in Population
New York City, New York	8,008,278	8,175,133	166,855 more people
Los Angeles, California	3,694,820	3,792,621	97,801 more people
Chicago, Illinois	2,896,016	2,695,598	200,418 fewer people
Houston, Texas	1,953,631	2,099,451	145,820 more people
Philadelphia, Pennsylvania	1,517,550	1,526,006	8,456 more people

1. Which city's population changed the most?

   _____

2. If Los Angeles continues to follow the same trend, how will its population change?

   _____

3. Do you think New York City's population will increase or decrease by 2020? Why?

   _____

   _____

   _____

Name: _____ **Date:**_____

Directions: Write a letter to a friend telling what you learned about cities. Use the Word Bank to help you complete your sentences.

Word Bank		
port	population	coast
inland	North America	cities

Dear _____ ,

   This week I learned about _____

_____

_____

   People are drawn to the coasts because _____

_____

_____

   If I could live anywhere, I would live _____

_____

                                   Your friend,

                          _____

Reading Maps

Name: _____ Date:_____

Directions: Study the map, and answer the questions.

The Great Basin

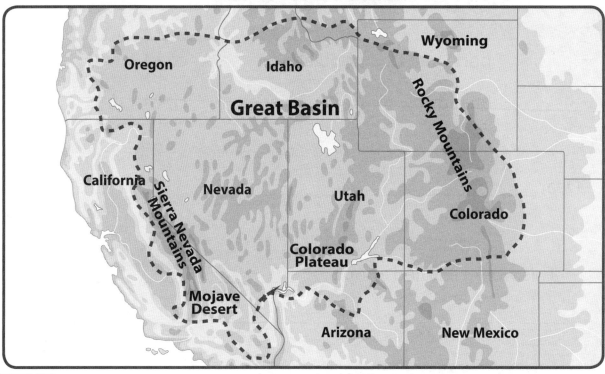

1. List at least three states that contain part of the Great Basin.

   _____

   _____

2. What two mountain ranges border the Great Basin?

   _____

   _____

3. What other physical features are part of the Great Basin?

   _____

   _____

Name: _____ **Date:**_____

Directions: Follow the steps to complete the map.

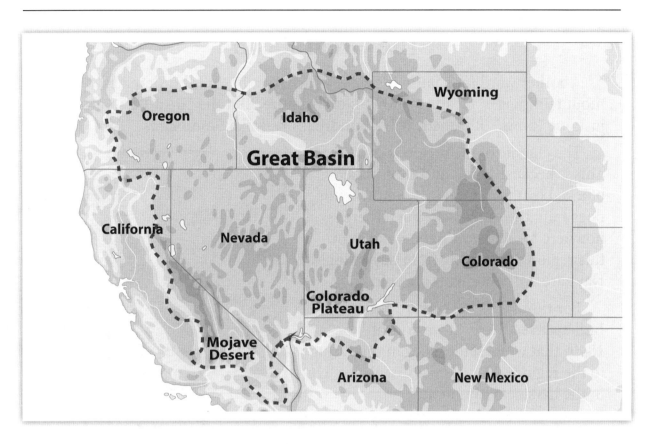

1. Add a compass rose and a title to the map.

2. Label the Rocky Mountains. They border the Great Basin to the east.

3. Label the Sierra Nevada Mountains. They border the Great Basin to the west.

4. Choose a color, and lightly shade the Great Basin region.

Challenge: Cover the state names with sticky notes. Try to name the states from memory. Write their names on top of the sticky notes. Then, lift the sticky notes to check your answers.

Read About It

Name: _____ **Date:** _____

Directions: Read the text, and study the photo. Then, answer the questions.

The Great Basin

Another word for a *basin* is a *bowl*. The Great Basin is bordered by mountain ranges, creating a bowl-like structure. But the Great Basin is not just one "bowl." It is made up of hundreds of similar landforms.

The climate of the region is quite dry. This is because winds and clouds are blocked by the mountains. But, when it does rain, the water stays put. There is nowhere for it to go! Temperatures in this region vary widely from day to night. Days are hot. Yet nights are cold.

For thousands of years, American Indians lived there. It wasn't until the 1840s that white settlers began migrating to the region. Many tribes learned to live in the harsh environment.

1. What does the word *basin* mean?

_____

2. Describe the climate of the Great Basin.

_____

_____

3. Why doesn't the Great Basin receive much rain?

_____

_____

Name: _____ Date:_____

Directions: Study the map closely. Then, answer the questions.

Tribes of the Great Basin

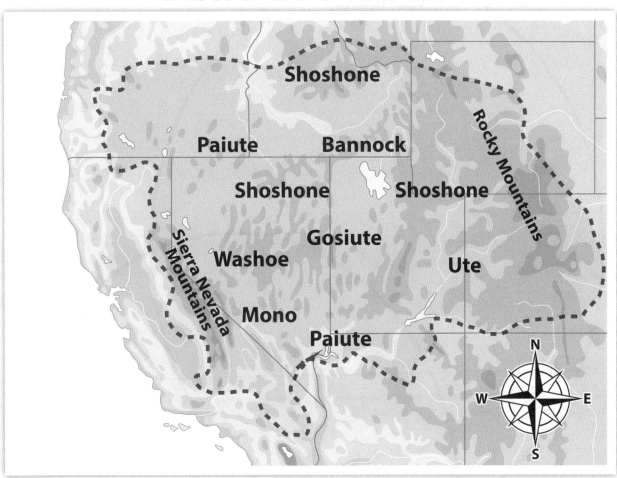

1. Which tribe is found throughout the northern Great Basin?

   _____

2. Which two tribes are closest to the Rocky Mountains in the east?

   _____

3. Which three tribes are the farthest west?

   _____

Name: _____ **Date:**_____

Directions: Compare and contrast your region to the Great Basin.

Geography and Me

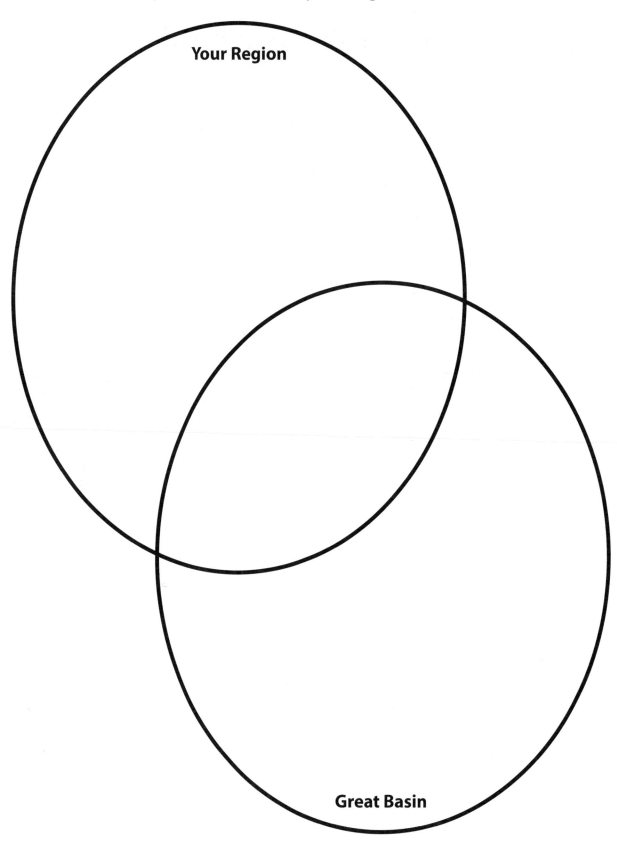

Your Region

Great Basin

© Shell Education

Name: _____ **Date:** _____

Directions: Study the diagram of the volcano. Then, answer the questions.

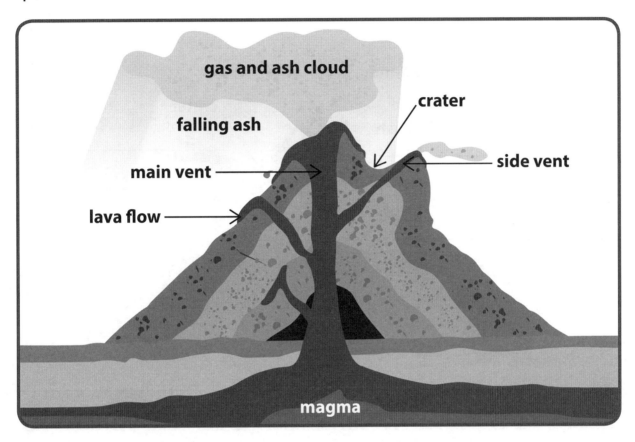

1. What is lava called when it is underground?

   _____

2. What is the difference between eruptions from the main vent and side vent?

   _____

   _____

3. What else erupts besides lava?

   _____

Creating Maps

Name: _____ **Date:** _____

Directions: Use the Word Bank to label the diagram. Then, describe what is happening in the diagram.

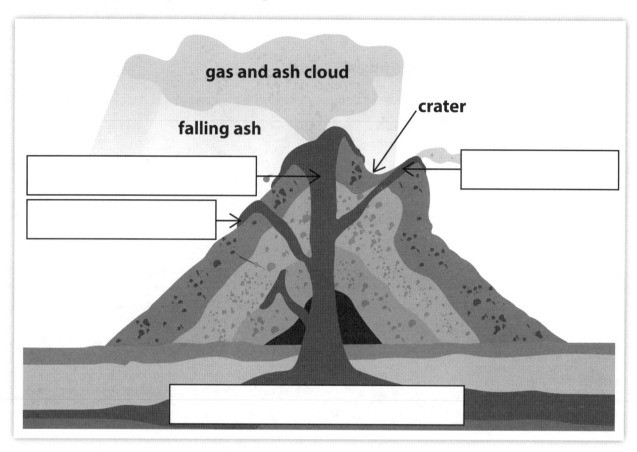

Word Bank	
main vent	side vent
lava flow	magma

_____

_____

_____

_____

28624—180 Days of Geography
© Shell Education

Name: _____ **Date:** _____

Directions: Read the text, and study the photo. Then, answer the questions.

A Park Unlike Any Other

There is a unique park on an island. It is called the Hawai'i Volcanoes National Park. The park is home to two active volcanoes. They are Mauna Loa and Kilauea. Visitors come from all over to see the two great landforms.

Both landforms are large shield volcanoes. These are volcanoes that look like domes. Mauna Loa is the largest in the world! But Kilauea has its own claim to fame. It is the most active in the world. As Kilauea erupts, lava spills over the land. Over time, the lava cools and becomes part of the island.

The Hawai'ian Islands were formed by this process. For millions of years, volcanoes erupted. Their lava flowed and then cooled. The islands grew slowly but surely. You can visit the park to see this process in action!

1. How does the photo compare to the text you read?

_____

_____

2. How have volcanoes shaped the Hawai'ian Islands?

_____

_____

3. Why is Mauna Loa famous?

_____

_____

Think About It

Name: _____ **Date:**_____

Directions: This diagram shows different types of volcanoes. Study the images closely. Then, answer the questions.

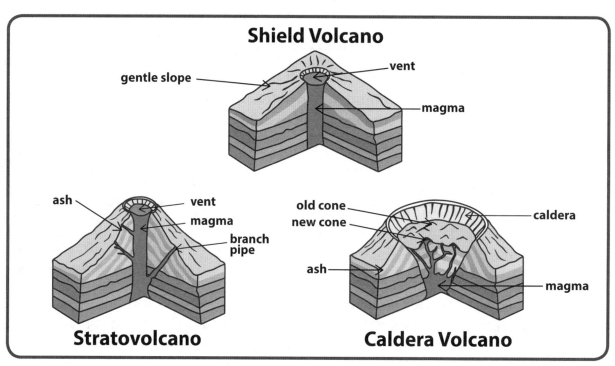

1. Describe a shield volcano.

   _____

   _____

2. How is the shield volcano different from the stratovolcano?

   _____

   _____

3. How is the stratovolcano different from the caldera volcano?

   _____

   _____

Name: _____ **Date:** _____

Directions: Share what you learned about volcanoes through social media posts. Write one fact per post. Remember to keep them short and sweet!

☐ ——————————————— 🐦

_____

_____

_____

☐ ——————————————— 🐦

_____

_____

_____

☐ ——————————————— 🐦

_____

_____

_____

Name: _____ **Date:** _____

Directions: There are many levels of government. Study the map closely. Then, answer the questions.

Reading Maps

Levels of Government

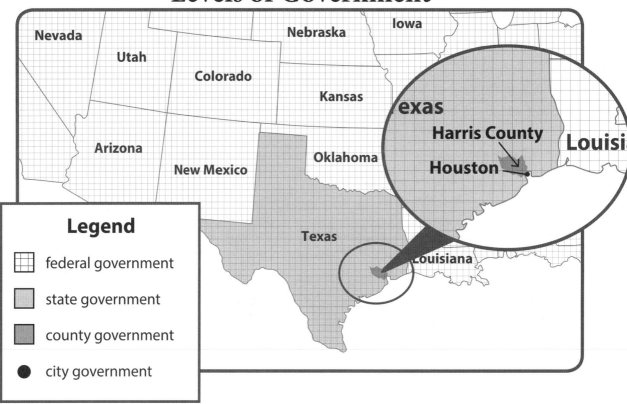

1. Texas is an example of what level of government?

   _____

2. Which level of government rules over all the states?

   _____

3. Which level of government is in charge of a large area or county? How do you know?

   _____

   _____

Name: _____ **Date:**_____

Directions: Follow the steps to complete the map.

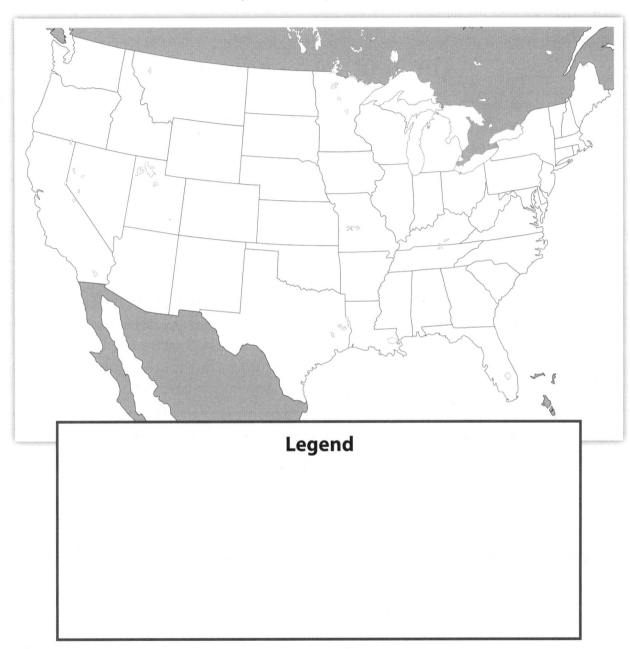

Legend

1. Lightly shade one state. This represents state government.

2. Lightly shade all the states (including the one you chose) in another color. This represents federal government.

3. Create a legend to show what each color represents.

Read About It

Name: _____ **Date:**_____

Directions: Read the text, and study the photo. Then, answer the questions.

The Tenth Amendment

Have you ever heard of the U.S. Constitution? It is the document that tells Americans their rights. It also tells how the country should be run. Many lives are shaped by its words. It was written in 1787. This was a long time ago! In the years since it was written, it has been changed, or amended. The first 10 amendments, or changes, are the Bill of Rights.

The Tenth Amendment is important. It talks about power. In the United States, there are levels of power. The federal government oversees the whole country. States focus on the needs of the people within their states. There are also local powers. These are cities and counties. The Tenth Amendment says that any powers not given to the federal government are given to the states. This means that the states have powers that the country does not.

1. What does the word *amend* mean?

_____

2. What are three levels of power in the United States?

_____

_____

3. Describe the Tenth amendment.

_____

_____

Name: _____ **Date:**_____

Directions: This diagram shows some of the federal, state, and local government powers. Study the diagram, and answer the questions.

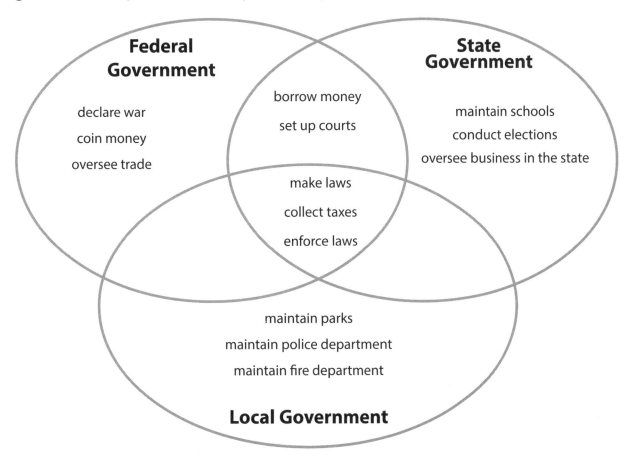

Federal Government

declare war

coin money

oversee trade

borrow money

set up courts

make laws

collect taxes

enforce laws

State Government

maintain schools

conduct elections

oversee business in the state

maintain parks

maintain police department

maintain fire department

Local Government

Think About It

1. Name one power only the federal government has. Why might the federal government have this power?

   _____

   _____

2. Name one power only local governments have. Why might local governments have this power?

   _____

   _____

Geography and Me

Name: _____ **Date:** _____

Directions: Draw places that are maintained by your local government.

Name: _____ **Date:** _____

Directions: This map shows private and public forests in the United States. Study the map closely. Then, answer the questions.

Public and Private Forests

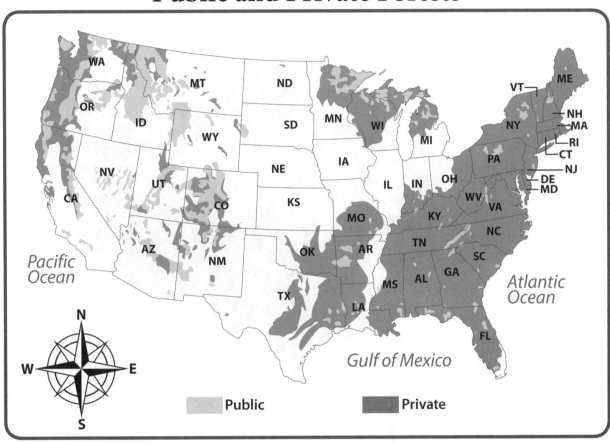

1. Where are there more private forests?

   _____

2. Where are there more public forests?

   _____

3. Name at least three states that do not have many private or public forests.

   _____

   _____

Creating Maps

Name: _____ **Date:**_____

Directions: Follow the steps to complete the map.

Public and Private Forests

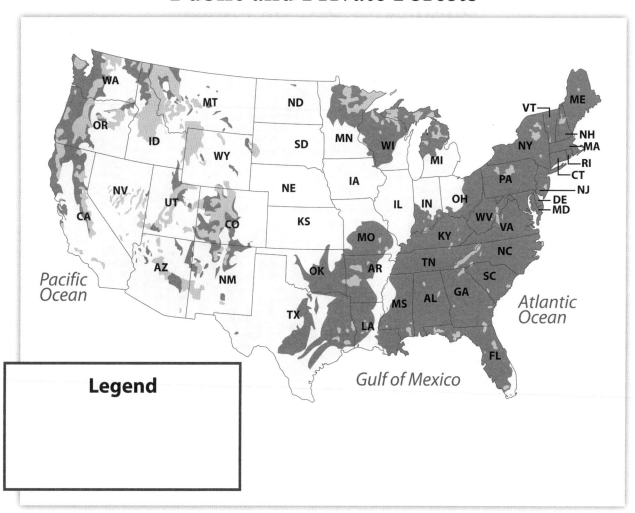

Legend

1. Create a legend to show the private and public forests in the United States. Private forests are dark gray, and public forests are light gray.

2. Circle the largest areas of public forests on the map.

3. Draw a box around the largest areas of private forests on the map.

4. Add the circle and box to your legend.

5. Add a compass rose.

Name: _____ **Date:** _____

Directions: Read the text, and study the photo. Then, answer the questions.

Forests of the United States

In the United States, people use trees for different purposes. In the South, most forests are used for timber. These trees are used to build things such as homes and other structures. Timber is used as a resource to create wealth in the South. For this reason, the South is called the "woodbasket" of the country. Forests in the North are also mainly used for timber. For the most part, forests in the North and South can be found on private land.

In the West, there are fewer forests. Many of them are public. That is why most national parks are in the West. Such parks are public property.

1. How are trees used in the North and the South?

_____

2. On what type of land are they usually found?

_____

3. In which region are most national parks found?

_____

4 . Why are both public and private forests important?

_____

_____

Name: _____ **Date:**_____

Think About It

Directions: The graphs show how much timber the United States uses. Study the graphs carefully. Then, answer the questions.

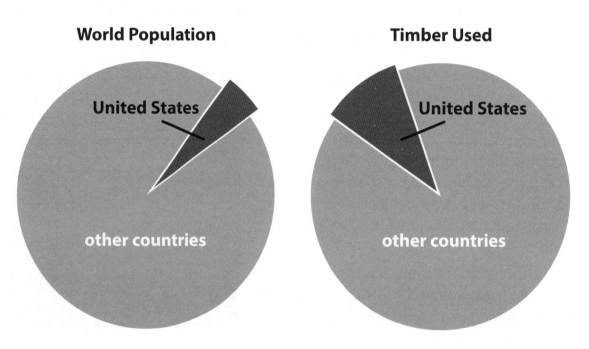

World Population

United States

other countries

Timber Used

United States

other countries

1. About how much of the world's population does the United States make up?

   _____

2. Compare the U.S. population to the amount of timber the country uses. Which is larger?

   _____

3. Do you think the United States uses more than its share of the world's timber? Why or why not?

   _____

   _____

Name: _____ **Date:** _____

Directions: Write a letter to a friend telling what you learned about forests. Use the Word Bank to help you complete your sentences.

Word Bank			
South	West	North	timber
resource	public forests	private forests	population

Dear _____,

This week, I learned about _____

_____

_____

There are different types of forests, _____

_____

_____

The United States uses _____

_____

_____

Your friend,

_____

Reading Maps

Name: _____ **Date:**_____

Directions: This map shows Oregon and the states around it. Study the map, and answer the questions.

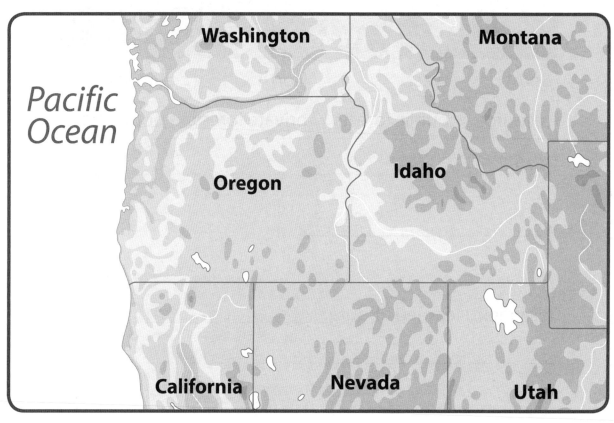

Washington

Montana

Pacific Ocean

Oregon

Idaho

California

Nevada

Utah

1. Which states border Oregon?

   _____

   _____

2. What ocean borders Oregon?

   _____

3. What opportunities do you think Oregon might have because of its position on the coast?

   _____

   _____

Name: _____ **Date:**_____

Directions: Use the clues to label the states and body of water that border Oregon.

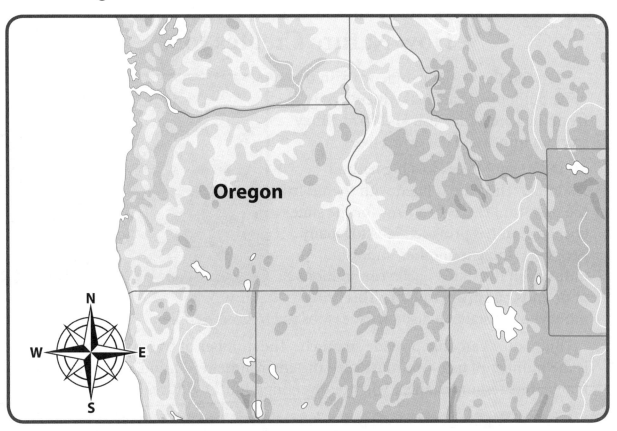

1. Idaho borders Oregon to the east.

2. California and Nevada border Oregon to the south.

3. California is west of Nevada.

4. Washington borders Oregon to the north.

5. The Pacific Ocean borders Oregon to the west.

Challenge: Cover the state names with sticky notes. Try to name the states from memory. Write their names on top of the sticky notes. Then, lift the sticky notes to check your answers.

Name: _____ Date:_____

Directions: Read the text, and study the photo. Then, answer the questions.

How Nature Creates Jobs

Around the world, people work in the great outdoors. They make their living from the land. They fish and farm. They work in mines and cut down trees.

In Oregon, trees are a big natural resource. The state is filled with them! In the past, timber has been key for the state's economy. It has provided jobs for many people. Fishing is another important industry in the state. Each year, millions of pounds of fish are caught. This brings money to the state. It also provides jobs to Oregon's people.

But nature can only provide so much. The number of trees is limited. So is the number of fish. We must keep the environment in mind. If we take too many of its resources, there will not be enough for the future.

1. How does nature create jobs?

_____

_____

2. What are two of Oregon's important natural resources?

_____

3. Why are natural resources limited?

_____

_____

Name: _____ Date:_____

Directions: This chart shows the number of mining and logging jobs each year. Study the graph, and answer the questions.

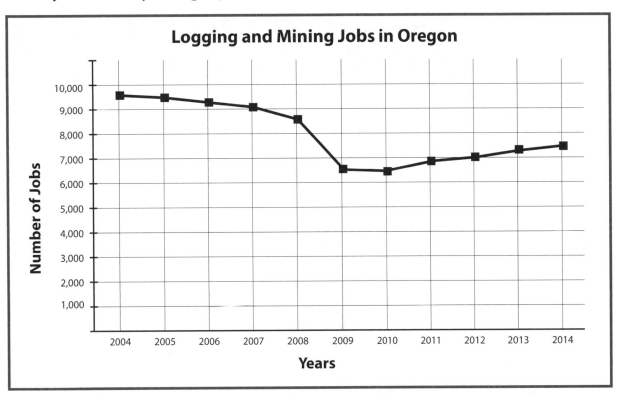

Logging and Mining Jobs in Oregon

Number of Jobs

Years

Think About It

1. In which year were there the most logging and mining jobs?

   _____

2. How many logging and mining jobs do you think there were in 2015? Why?

   _____

   _____

3. If you were looking for a stable job, would you choose a career in logging or mining? Why or why not?

   _____

   _____

Geography and Me

Name: _____ **Date:** _____

Directions: Write five jobs that are created by natural resources. Then, answer the questions. An example has been done for you.

Job	Natural Resource
wheat farmer	*soil*

1. Which of these natural resources are available in your area?

   _____

   _____

2. Why are natural resources important?

   _____

   _____

Name: _____ Date:_____

Directions: In 1838, the Cherokee people were forced to leave their homes and walk west. The map shows two of their routes. Study the map, and answer the questions.

The Trail of Tears

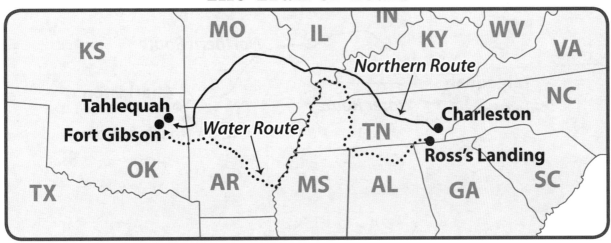

1. In which state do both routes begin?

   _____

2. How many states did the Cherokee have to cross on the Northern Route? List the states.

   _____

   _____

3. Which route is a longer distance? How do you know?

   _____

   _____

   _____

   _____

Creating Maps

Name: _____ Date:_____

Directions: This map shows two of the routes on the Trail of Tears. Follow the steps to complete the map.

The Trail of Tears

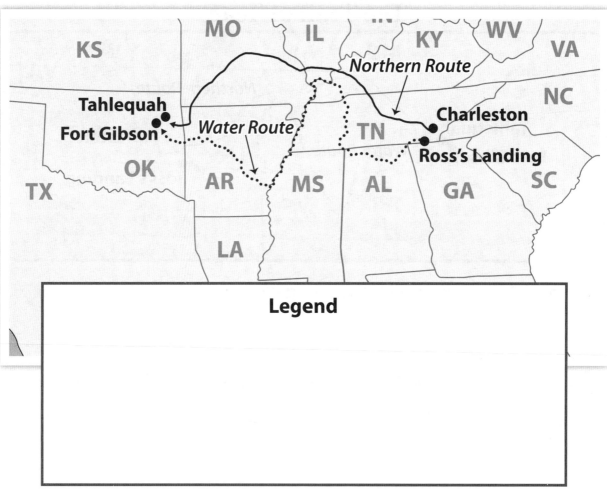

Legend

1. Trace the northern route in red.

2. Trace the water route in blue.

3. Outline the state borders in green.

4. Create a legend to show what the colors mean.

Challenge: Cover all of the state abbreviations. Name as many states as you can. Then, check your answers.

Name: _____ **Date:** _____

Directions: Read the text. Then, answer the questions.

Trail of Tears

In 1829, gold was discovered in Georgia. The gold was on Cherokee land, but white people wanted it. They wanted to mine the land and make money. They also wanted to use the land for its other resources. The soil was good for farming. White people thought the United States should own tribal land.

Sadly, President Jackson agreed. He also wanted the land for the United States. As a result, Congress passed the Indian Removal Act. This act said that the president could make the tribes move west. Some tribes agreed to move. But the Cherokee did not. It was their land and their home. They fought back in court. But the lawsuits did not save their land.

In 1838, the United States forced the Cherokee to leave their homes. They were sent to prison camps to wait for orders. After weeks at the camps, they began the Trail of Tears. The Cherokee walked hundreds of miles. Many of them were on foot. Thousands died during the journey. By 1839, the last Cherokee people had been forced from the land.

1. Why did the United States want the Cherokee land?

_____

2. Who supported the Indian Removal Act?

_____

_____

3. The Trail of Tears is an example of involuntary migration. How might this be different from when people choose to move?

_____

_____

Think About It

Name: _____ **Date:** _____

Directions: Study the time line. Then, answer the questions.

1829—Gold is found in Georgia on Cherokee Land.

1831—The Cherokee bring their case to court.

1838—The United States forces the Cherokee to move to prison camps (mostly in Tennessee).

1838—The Cherokee walk from Tennessee to Oklahoma.

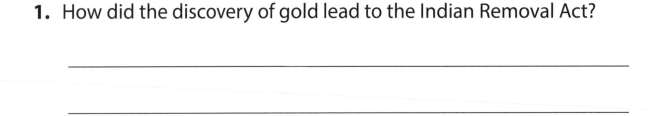

1830   1835   1840

1830—The Indian Removal Act is passed.

1839—The last group of Cherokee arrives in Oklahoma.

1. How did the discovery of gold lead to the Indian Removal Act?

_____

_____

_____

2. How did the Cherokee fight back against the Indian Removal Act?

_____

_____

_____

3. In what year did the Cherokee's involuntary migration begin?

_____

_____

Name: _____ **Date:** _____

Directions: Think about involuntary migration. Why is it wrong? List reasons on the diagram.

Involuntary migration is wrong.

Reading Maps

Name: _____ **Date:** _____

Directions: Study the map closely. Then, answer the questions.

Canadian Provinces

1. List three cities found in the Quebec province.

   _____

2. What do you notice about the location of the cities in Quebec?

   _____

3. In Quebec, do you think more people live near or far from the water? How do you know?

   _____

   _____

Name: _____ **Date:** _____

Directions: Follow the steps to complete the map.

Canadian Provinces

1. Add a compass rose to the map.

2. Draw a box around the three Canadian territories that are the farthest north.

3. Choose another color, and shade the three Canadian provinces that are the farthest west.

4. Choose a third color, and shade the remaining provinces.

5. Outline province borders in black.

Challenge: Cover the map, and name as many Canadian provinces as you can. Then, uncover the map to check your answers.

Read About It

Name: _____ **Date:**_____

Directions: Read the text, and study the photo. Then, answer the questions.

Same Country, Different World

Quebec is a province in Canada. But it is different from other provinces. Many people speak French in Quebec. In fact, most people who live there say French is their first language. In Canada's other provinces, most people's first language is English.

There is a reason for this difference. Quebec was founded by the French. In 1534, French explorers claimed the land. They named it New France. Later, the British gained control of the region. But many people in Quebec have ancestors who lived in New France.

Quebec's history and language define its culture. It affects what people eat and what they wear. It even affects what they watch on TV! For example, the most watched shows in Quebec are French shows. And most schools in Quebec teach in French. Street signs and business signs are also French. For these reasons and others, Quebec is unique. Even though it is part of Canada, Quebec is its own world.

1. Why does Quebec have so many French speakers?

_____

_____

2. How is Quebec's culture different from the rest of Canada?

_____

_____

3. How do you think language influences culture?

_____

_____

Name: _____ **Date:**_____

Directions: This circle graph shows Canadians' first languages. Study the graph carefully. Then, answer the following questions.

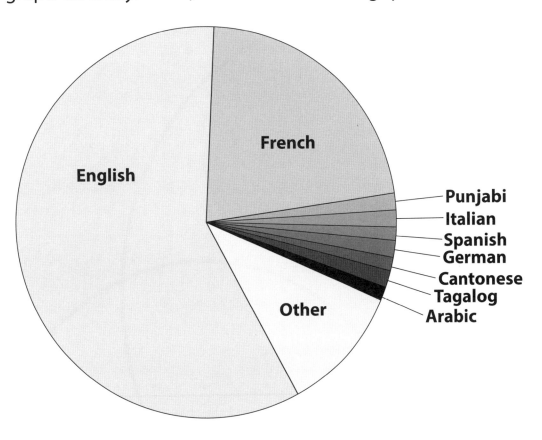

1. What are the top two first languages in Canada?

   _____

2. Name three less common languages in Canada.

   _____

3. How does this graph show that different cultures can share the same country?

   _____

   _____

Think About It

Geography and Me

Name: _____ **Date:** _____

Directions: Compare and contrast where you live to Quebec.

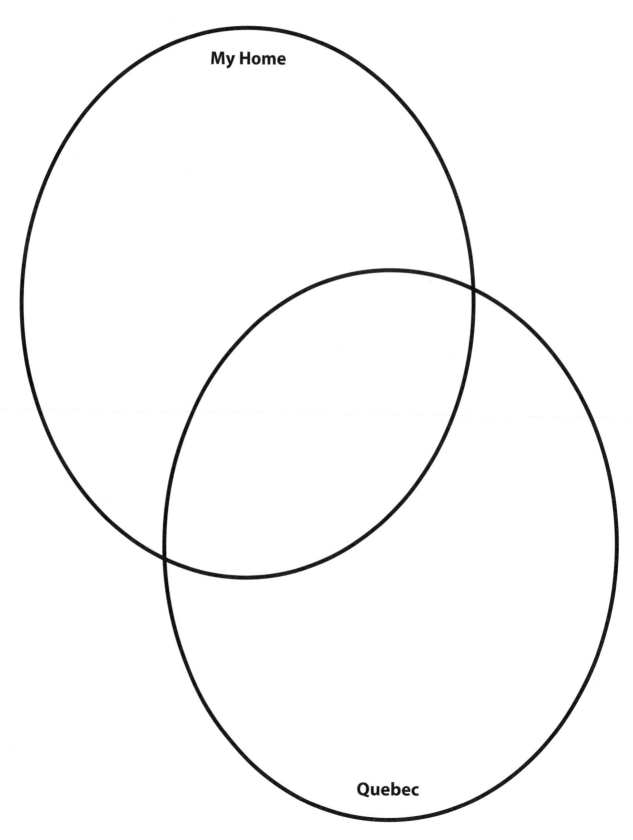

My Home

Quebec

Name: _____ **Date:**_____

Directions: In southern Florida, there is a large marsh called the Everglades. Study the map, and answer the questions.

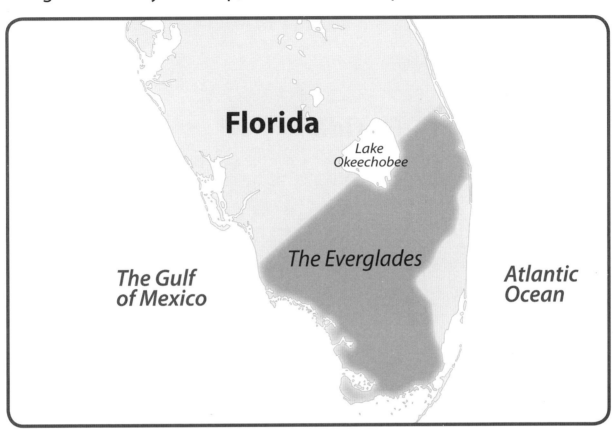

1. The Everglades is bordered by what bodies of water?

   _____

   _____

2. What is the large lake that is part of the Everglades environment?

   _____

3. Describe what you think a marsh looks like.

   _____

   _____

Creating Maps

Name: _____ **Date:** _____

Directions: Follow the steps to complete the map.

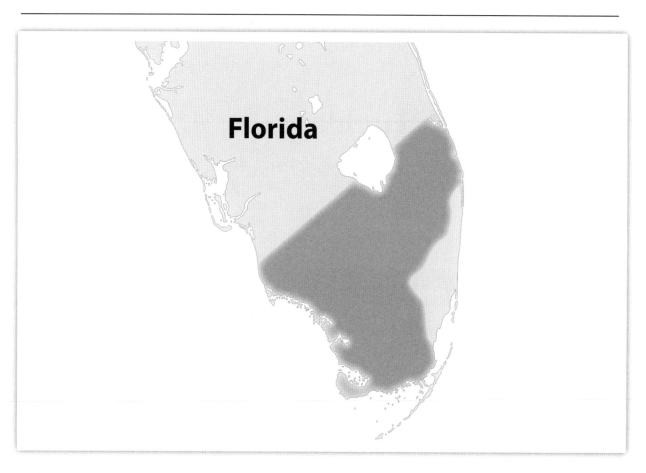

Florida

1. Label the Everglades.

2. Outline the Everglades in green.

3. Label Lake Okeechobee.

4. Label the Atlantic Ocean.

5. Label the Gulf of Mexico.

6. Add a compass rose to the map.

7. Add a title to the map.

28624—180 Days of Geography   © *Shell Education*

Name: _____ **Date:**_____

Directions: Read the text, and study the photo. Then, answer the questions.

The Wild of the Wetlands

The Everglades is bursting with life. Many plants and animals call this place home. The region is a wetland. It has marshes and swamps. The water that covers the land is shallow. It is usually not more than one foot deep. One of the most common plants in the region is sawgrass. It is all over! Sawgrass can grow up to 10 feet in length.

Many creatures live in these wetlands. There are turtles and snakes. There are bobcats and otters. The Everglades is also known for its many types of birds. Endangered animals include manatees and crocodiles.

The Everglades was in serious danger not too long ago. In the 1800s, people were destroying the land quickly. They wanted to turn the marsh into farmland. To do this, they drained as much of the marsh as they could. Today, only half of the original wetlands still exists.

1. What is a wetland?

   _____

2. Describe the plants and animals in the Everglades.

   _____

   _____

3. Why were the Everglades being destroyed in the 1800s?

   _____

   _____

Think About It

Name: _____ **Date:** _____

Directions: This chart shows the average rainfall in the Everglades. Study the chart, and answer the questions.

Month	Rainfall (inches)	Month	Rainfall (inches)
January	1.65	July	7.07
February	1.85	August	8.30
March	1.92	September	8.71
April	2.77	October	5.54
May	5.86	November	2.28
June	9.07	December	1.37

1. Which month is the rainiest? Which is the driest?

   _____

   _____

2. Which months do you think are the rainy season? Why?

   _____

   _____

3. How might this pattern affect the animals in the Everglades?

   _____

   _____

Challenge: Use the charts to create a bar graph on a separate sheet of paper. Don't forget to include a title and labels.

Name: _____ **Date:**_____

Directions: Imagine you are in the Everglades. What plants and animals do you see? Draw a picture of the Everglades ecosystem.

Reading Maps

Name: _____ **Date:** _____

Directions: A biome is a region that has similar ecosystems and climates. Study the biome map, and answer the questions.

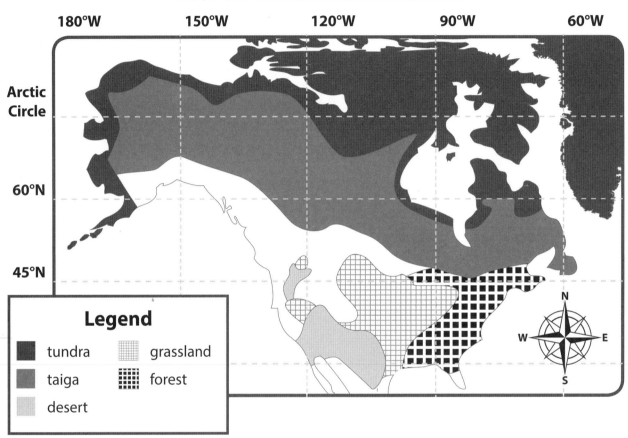

North American Biomes

Legend

- tundra
- taiga
- desert
- grassland
- forest

1. The taiga biome is south of which biome?

_____

2. Which biomes are south of the taiga?

_____

_____

3. The coordinates 60°N, 120°W fall within which biome?

_____

Name: _____ **Date:** _____

Directions: Follow the steps to complete the map.

North American Biomes

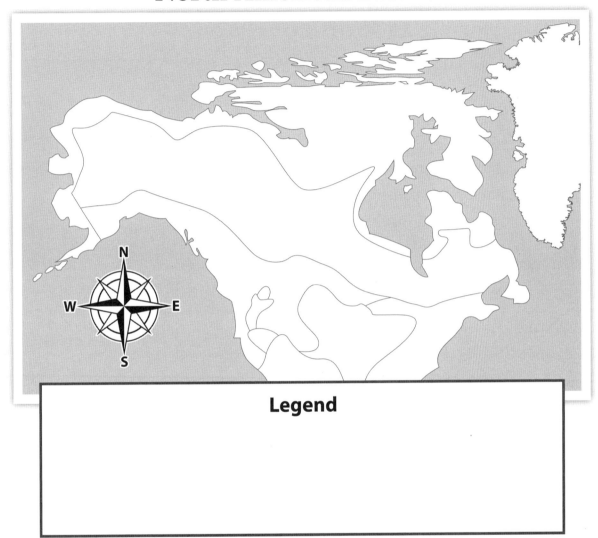

1. Choose a color, and use it to shade the tundra biome. It is the farthest north.

2. Choose a different color, and use it to shade the taiga biome. It is directly south of the tundra biome.

3. Complete the legend to show the colors you used.

4. Add a question mark on any biome that you would like to know more about.

Read About It

Name: _____ **Date:** _____

Directions: Read the text, and study the photo. Then, answer the questions.

A World of Green

The taiga is covered in trees. It is a world of green. The most common types of trees in this biome are conifers. Most conifer trees have needle-like leaves and grow cones. Some examples include firs, pines, and spruces. When you think of a Christmas tree, it is likely a conifer that pops into your mind.

These trees are well suited to the taiga biome, a place famous for being cold. Winters are long. Summers are short and cool. So when the sun is out, these trees absorb as much sunlight as possible. Their branches are angled toward the ground. This means that when snow falls, it falls off the trees and onto the ground.

But the world of the taiga is not just green. The land is filled with lakes and rivers, as well. Some creatures that live in the biome are wolves, squirrels, lynxes, and moose. There are even reindeer!

1. Describe the climate in the taiga.

_____

_____

2. Why are conifers well suited for the taiga?

_____

_____

3. What are some animals that live in the taiga?

_____

Name: _____ Date:_____

Directions: This graph shows the populations of the lynx and snowshoe hare. The lynx preys on the snowshoe hare. Both animals live in the taiga. Study the graph, and answer the questions.

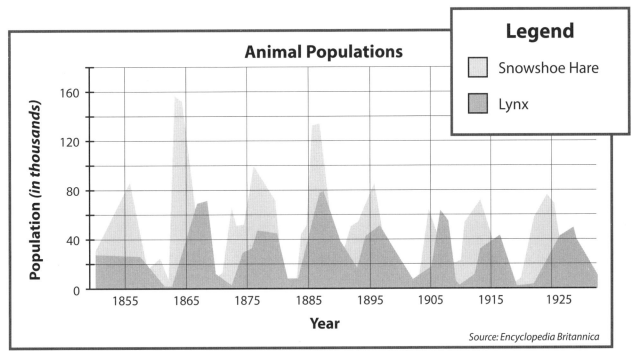

1. In which two years were there the most snowshoe hares?

   _____

2. In which two years were there the most lynx?

   _____

3. How are the populations of the snowshoe hare and the lynx connected?

   _____

   _____

   _____

Think About It

Geography and Me

Name: _____ **Date:** _____

Directions: Draw an animal that lives in the taiga biome. Describe the taiga biome below your drawing.

_____

_____

_____

_____

_____

_____

Name: _____ Date:_____

Directions: This map shows the groups that controlled North America in 1750. Study the map closely. Then, answer the questions.

North America in 1750

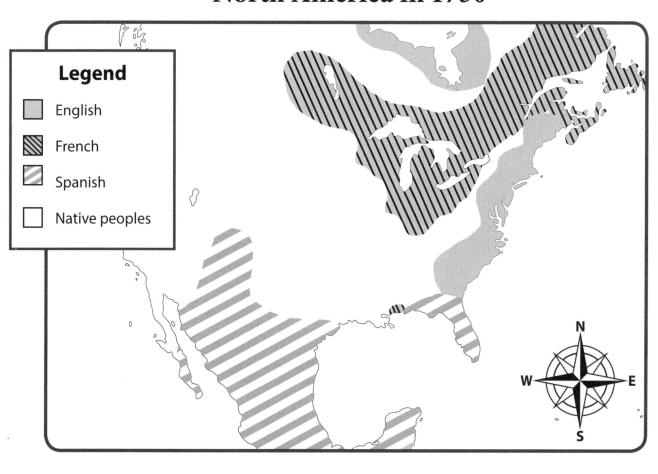

Legend

English

French

Spanish

Native peoples

1. Which group controlled most of North America at this time?

   _____

2. Which three European powers shared the East Coast?

   _____

   _____

   _____

Reading Maps

Creating Maps

Name: _____ **Date:** _____

Directions: Follow the steps to complete the map.

North America in 1750

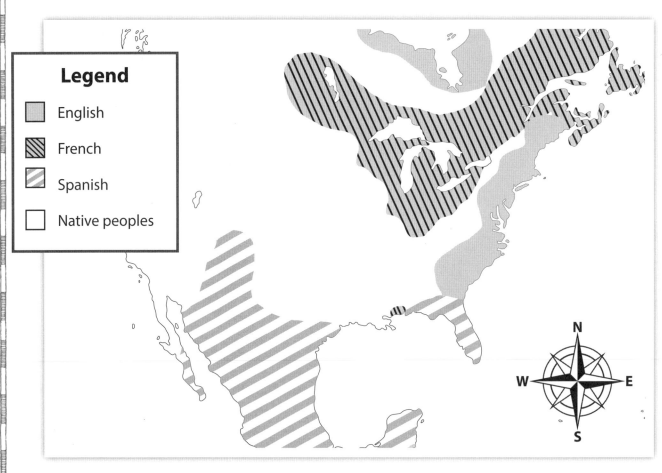

Legend

English

French

Spanish

Native peoples

1. Outline the French territory in blue.

2. Outline the English territory in red.

3. Outline the Spanish territory in green.

4. Use a different color to shade American Indian land.

5. Update the legend to include your outlines and shading.

Name: _____ Date:_____

Directions: Read the text, and study the photo. Then, answer the questions.

Making America

In 1776, the United States declared its independence. It would no longer serve Great Britain. The next hundred years brought a lot of change. The country grew and grew. It expanded west. In time, the Land of the Free reached the Pacific Ocean.

One of the biggest changes was in 1803. This was the year of the Louisiana Purchase. The United States bought a vast amount of land from France. The country was now twice as big! The new land started at the Mississippi River. It continued until the Rocky Mountains.

This land was bought from Napoleon. He was a French ruler. He wanted to sell the land so that he could fund future wars. The United States only paid $15 million for the land!

1. What was the Louisiana Purchase?

   _____

   _____

2. From whom did the United States buy the land?

   _____

3. Why do you think the author stated that the United States "only paid $15 million?"

   _____

   _____

Read About It

Name: _____ **Date:** _____

Directions: Compare the two maps below. The top map shows the United States in 1854. The bottom map shows the United States today. Study the maps, and answer the questions.

Think About It

1. How are the maps different?

   _____

   _____

   _____

   _____

   _____

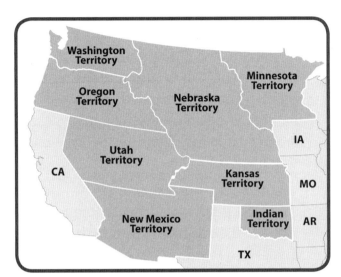

2. What has the Nebraska Territory become since 1854?

   _____

   _____

   _____

   _____

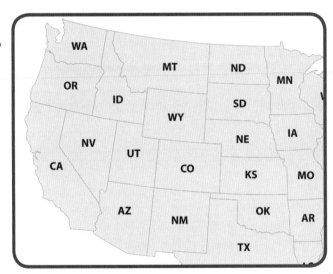

3. How was the West changed since 1854?

   _____

   _____

   _____

Name: _____ **Date:** _____

Directions: How has North America changed since 1750? List examples on the chart.

North America in 1750	North America Today

Geography and Me

Reading Maps

Name: _____ **Date:** _____

Directions: Study the map, and answer the questions.

Yellowstone National Park

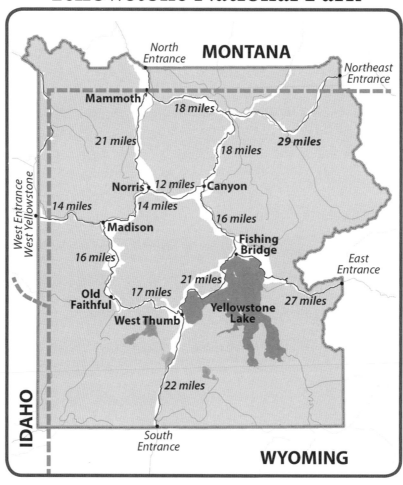

1. In what state is most of the park found?

_____

2. How far is it from Mammoth to Norris?

_____

3. How far is it from West Yellowstone to Old Faithful?

_____

Name: _____   Date:_____

Directions: Follow the steps to complete the map.

Yellowstone National Park

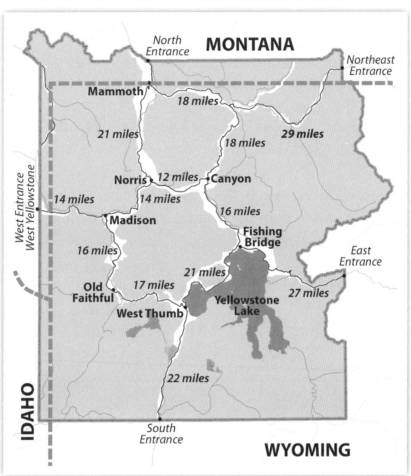

1. Trace a route from North Entrance to West Thumb. How many miles is your route?

   _____

2. Trace a route from Northeast Entrance to Old Faithful. How many miles is your route?

   _____

3. Which two points on the map are the closest together?

   _____

Read About It

Name: _____ **Date:** _____

Directions: Read the text, and study the photo. Then, answer the questions.

National Treasure

There is so much to see and do at Yellowstone National Park. The park is most famous for a geyser called Old Faithful. It erupts regularly, every half hour to two hours. Water sprays 100 feet in the air. People love to take pictures of Old Faithful when it erupts.

Yellowstone also has thousands of hot springs. These are waters warmed by magma below Earth's surface. If you visit the park, you can soak in the Boiling River. It is filled with runoff from the hot springs.

Yellowstone Lake is filled with activity. People take boat tours and learn about the wildlife. Yellowstone also has a museum. You can learn about the park's history there. These are just a few things to do at the park. There are many more!

1. How do you think Old Faithful got its name?

   _____

   _____

2. What are two things a visitor might do while at the park?

   _____

   _____

3. If you were interested in learning about how the park got started, where would you go?

   _____

Name: _____ **Date:** _____

Directions: These photos show different things to do in Yellowstone. Study the photos, and read the captions. Then, answer the questions.

People watch fish in Yellowstone Lake from Fishing Bridge.

People can take canoes on Yellowstone Lake.

1. Describe what a person might see from Fishing Bridge.

   _____

   _____

2. Describe what a person might see on a boat tour.

   _____

   _____

3. How can two people have different experiences at Yellowstone Lake?

   _____

   _____

   _____

Think About It

Name: _____ **Date:**_____

Directions: Share what you learned about Yellowstone National Park through social media posts. Write one fact per post. Remember to keep them short and sweet!

Geography and Me

[ ] _____ 🐦

_____

_____

_____

[ ] _____ 🐦

_____

_____

_____

[ ] _____ 🐦

_____

_____

_____

Name: _____ **Date:** _____

Directions: This is a map of the Caribbean. Study the map closely. Then, answer the questions.

1. Which country is nearest 20°N and 70°W?

   _____

2. List three water bodies that surround the Caribbean.

   _____

   _____

3. List the two continents closest to the Caribbean.

   _____

   _____

Creating Maps

Name: _____ **Date:** _____

Directions: Follow the steps, and answer the questions.

1. Circle the islands labeled *Greater Antilles*.

2. Put a square around the islands labeled *Lesser Antilles*.

3. Why do you think the islands on the right are called *Greater* Antilles and the islands on the left are called *Lesser* Antilles?

   _____

   _____

4. These islands are in the tropics. What sorts of goods do you think they sell to other nations?

   _____

   _____

Name: _____ **Date:** _____

Directions: Read the text, and study the photo. Then, answer the questions.

The Life Aquatic

Life near the Caribbean Sea can feel like paradise. That is why so many people like to take trips there! The islands are very popular. They have clear waters. They also have diverse plants and animals. And the weather is warm all year!

It is because of the climate that farmers are able to harvest year-round. This means they can grow more sugar, bananas, and coffee. These are some of the big exports from the islands. Exports are goods that are made in one country and shipped to other countries. This is how people in the Caribbean make money!

But Caribbean countries also spend a lot of money on imports. These are goods that are bought from other countries. In fact, the Caribbean tends to spend much more money on imports than exports. This is not good for the economy. It is never good to spend more than you make.

1. Why do people like to visit the Caribbean?

_____

_____

2. What are some of the Caribbean's main exports?

_____

3. Why are exports better than imports for an economy?

_____

_____

Read About It

Think About It

Name: _____ Date:_____

Directions: This bar graph shows trade between the Caribbean and Europe. Study the graph carefully. Then, answer the questions.

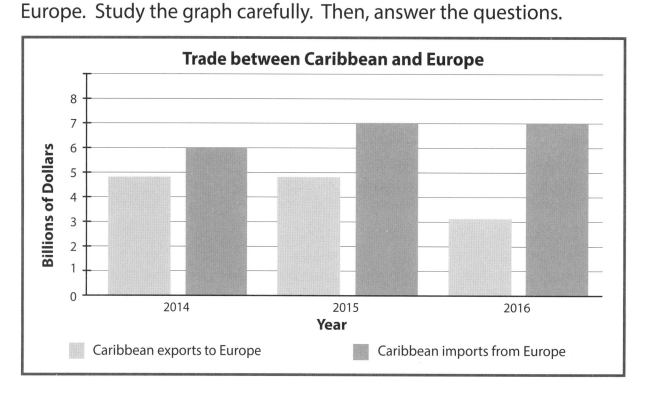

1. In which year did the Caribbean make the least money from exports?

   _____

2. In which year did the Caribbean spend the least money on imports?

   _____

3. Does the Caribbean import or export more goods? How do you know?

   _____

   _____

4. Did the Caribbean spend more money or make more money in the years shown?

   _____

Name: _____ **Date:** _____

Directions: Write a letter to a friend telling what you learned about the Caribbean. Use the Word Bank to help you complete your sentences.

Word Bank		
Greater Antilles	Lesser Antilles	Caribbean Sea
import	export	trade

Dear _____,

This week, I learned about _____

_____

_____

_____

The Caribbean is _____

_____

_____

It trades _____

_____

Your friend,

_____

Name: _____ **Date:** _____

Directions: Since women in the United States entered the workplace, they have made less money than men. This map shows how much the average woman earns for every dollar a man earns.

Wage Gap in Each State

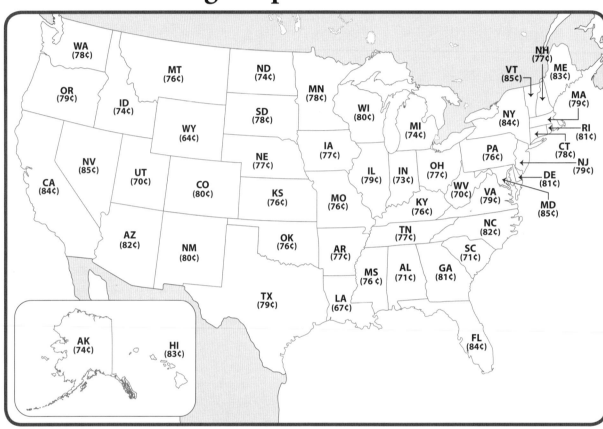

1. Which four states do women make the least money compared to men?

   _____

   _____

2. Which three states do women make the most money compared to men?

   _____

   _____

Reading Maps

Name: _____ **Date:**_____

Directions: This map shows how much women in the United States earn for every dollar a man earns. Follow the steps to complete the map.

Wage Gap in Each State

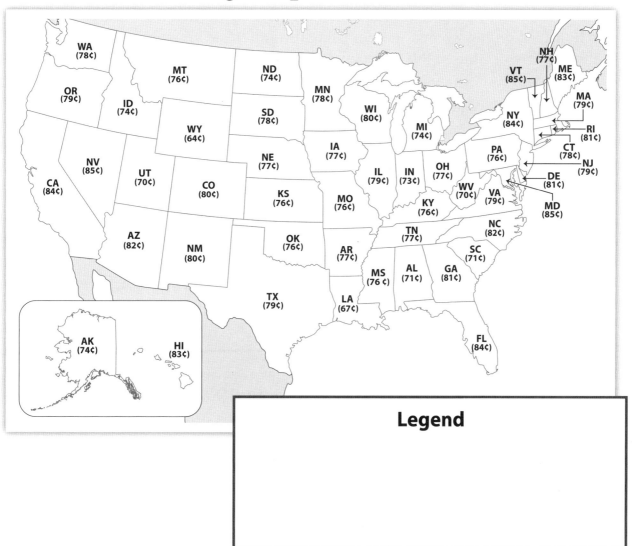

WA (78¢)	MT (76¢)	ND (74¢)	MN (78¢)		NH (77¢)	ME (83¢)	VT (85¢)

Legend

1. Shade the states where women make more than 80 cents on the dollar.

2. Use a different color to shade states where women make between 75 and 80 cents on the dollar.

3. Use a third color to shade the remaining states.

4. Create a legend that explains the different map colors.

Name: _____ Date:_____

Directions: Read the text, and answer the questions.

Read About It

Women Marching Forward

Women's roles have changed. In the early 1900s, women were told to stay at home. Working women were looked down upon. Women were not even allowed to vote! But in the last century, women have been marching forward, step by step.

When World War I began, more women entered the workforce. They needed to fill the jobs men had left when they went to war. Women were working. But they still were not granted the right to vote. It wasn't until 1920 that women finally got the right to vote. This was a huge step forward.

During World War II, many women entered the workforce. But after the war ended, they were expected to give their jobs back to men. Working women were still looked down upon in the 1950s. Despite this, more women entered the workforce.

In 2015, women were almost half of the U.S. workforce. But women are still not paid as much as men for the same jobs. This is called a *wage gap*. The women's movement is moving forward. But there is still much work to be done!

1. What was the role of women in the early 1900s?

_____

2. What happened to the workforce in World Wars I and II?

_____

_____

3. How has the women's movement marched forward?

_____

_____

Name: _____ Date:_____

Directions: This chart shows how much average American men and women make each year. Study the chart closely. Then, answer the questions.

1. About how much less money were women making than men in 2000?

    _____

2. About how much less money were women making than men in 2010?

    _____

3. Has the wage gap increased or decreased over time? How do you know?

    _____

    _____

    _____

Geography and Me

Name: _____ **Date:** _____

Directions: Think about working women in the United States. What steps need to be taken to create equal workplaces? List five steps on the diagram.

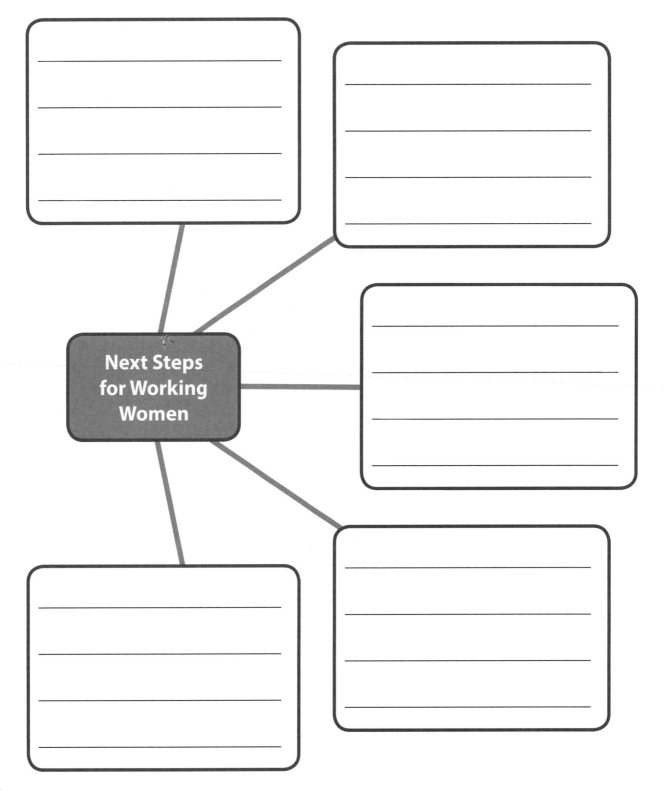

Next Steps for Working Women

Name: _____ **Date:** _____

Directions: On April 18, 1775, Paul Revere and other riders traveled to alert people of British attack. Study the map of his ride, and answer the questions.

Paul Revere's Ride

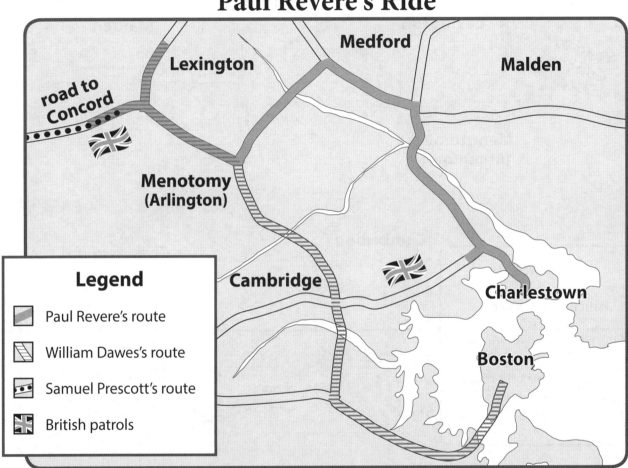

1. What cities were along Paul Revere's route?

   _____

2. How many times did Paul Revere's route go near British patrols?

   _____

3. Where do the routes of William Dawes and Paul Revere meet?

   _____

Creating Maps

Name: _____   Date: _____

Directions: Follow the steps to complete the map.

Paul Revere's Ride

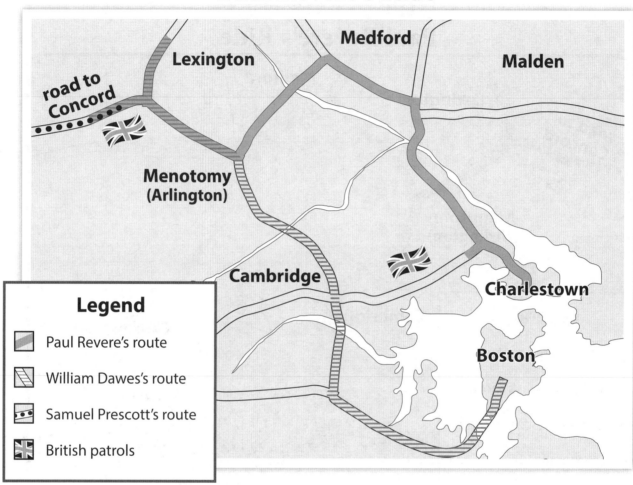

1. Trace each person's route in a different color.

2. Color the water blue.

3. Add the colors you used to the legend.

4. Place stars where two or more routes meet.

5. Circle the British patrols on the map.

Name: _____ **Date:**_____

Directions: Read the text, and study the picture. Then, answer the questions.

Paul Revere's Midnight Ride

Paul Revere could ride fast. He was an express rider. He carried news throughout the British colonies. But Revere never rode faster than he did on the night of April 18, 1775.

The day had come. The British were coming. They wanted to take the Patriots' weapons. The Patriots were people rebelling from British rule. Paul Revere was a proud Patriot.

On April 18, he and others rode to warn people about the British. Revere was almost captured at the start of his ride. But he escaped to ride through Medford and on to Lexington. Along the way, Revere told people to get ready. They were about to fight.

Revere continued his journey. He wanted to make sure that weapons in Concord had been hidden. But Revere was captured on his way to Concord. Still, his ride was a success. He had warned the Patriots. They were ready for the British!

1. Who was Paul Revere?

_____

2. Why did he ride on April 18, 1775?

_____

3. Was Paul Revere's ride successful? Why, or why not?

_____

_____

Think About It

Name: _____ Date:_____

Directions: Study the map, and answer the questions.

Routes of the British and Paul Revere

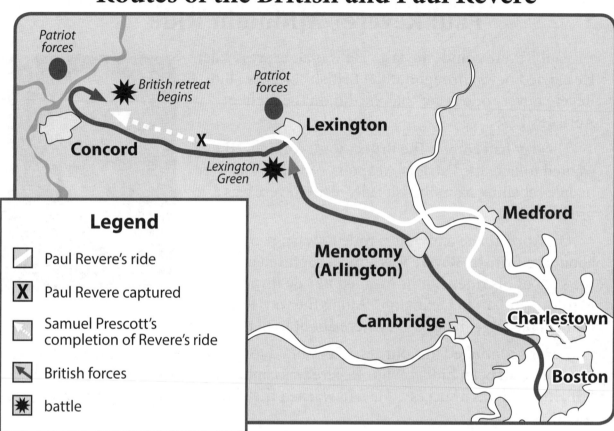

1. Where did the battles between the British and the Patriots take place?

   _____

2. Judging from the map, where did the Patriot forces win the battle?

   _____

3. What might have happened if Paul Revere had not made his famous ride?

   _____

   _____

Name: _____ **Date:** _____

Directions: Imagine you are Paul Revere. What do you see on your midnight ride? Draw a picture, and write a paragraph describing your journey.

_____

_____

_____

_____

_____

_____

Reading Maps

Name: _____ **Date:**_____

Directions: This diagram shows landforms that are often found in and around canyons. Study the diagram, and answer the questions.

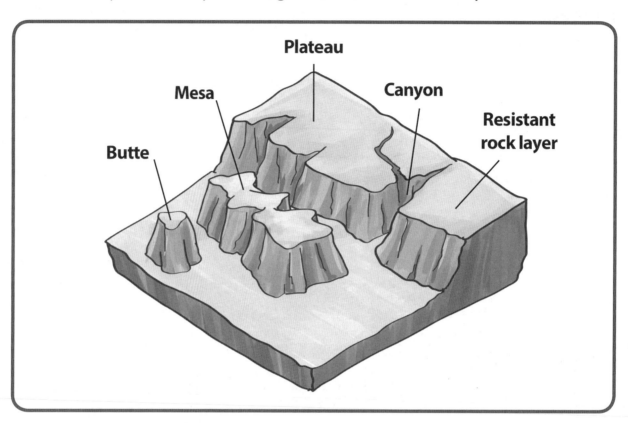

1. What do the tops of these landforms have in common?

   _____

2. How is a plateau different from a mesa?

   _____

   _____

3. Describe a canyon as shown in the diagram.

   _____

   _____

   _____

Name: _____   **Date:** _____

Directions: Create a diagram of landforms. Use the words in the Word Bank to label the photo. Some words may be used more than once.

Word Bank		
mesa	butte	plateau

Creating Maps

Read About It

Name: _____ **Date:**_____

Directions: Read the text, and study the photo. Then, answer the questions.

Carved by Water

The Grand Canyon is a stunning sight. Thousands of visitors come here each year to take in the views. When they arrive, they are actually at the top of the canyon. This is also called the *rim*. From the rim, visitors can see down into the canyon a mile deep.

At the bottom of the canyon is a river. It looks small from the rim. Yet this river formed the canyon over millions of years. As the river rushed over the rocks, it slowly broke off and moved small pieces of rocks. This is called *weathering* and *erosion*. Steep walls were formed as the river sank lower and lower. It is still carving away at the bottom of the canyon, little by little.

1. How was the Grand Canyon formed?

_____

_____

2. Why is the river at the bottom of the canyon?

_____

_____

3. Circle the rim of the canyon in the photo.

Name: _____ **Date:**_____

Directions: This chart shows the number of people who have visited the Grand Canyon over time. Study the chart, and answer the questions.

Year	Number of Visitors
1925	134,053
1935	206,018
1945	169,960
1955	892,400
1965	1,689,200
1975	2,625,100
1985	2,711,529
1995	4,557,645
2005	4,401,522
2015	5,520,736

1. Draw green circles next to the years in which the number of visitors increased.

2. When did the number of visitors decrease?

   _____

3. Do you think the number of visitors will increase or decrease in 2025? Why?

   _____

   _____

   _____

Name: _____ **Date:** _____

Directions: Write a type of landform that is near you. Complete the Venn diagram to compare and contrast that landform with canyons.

Geography and Me

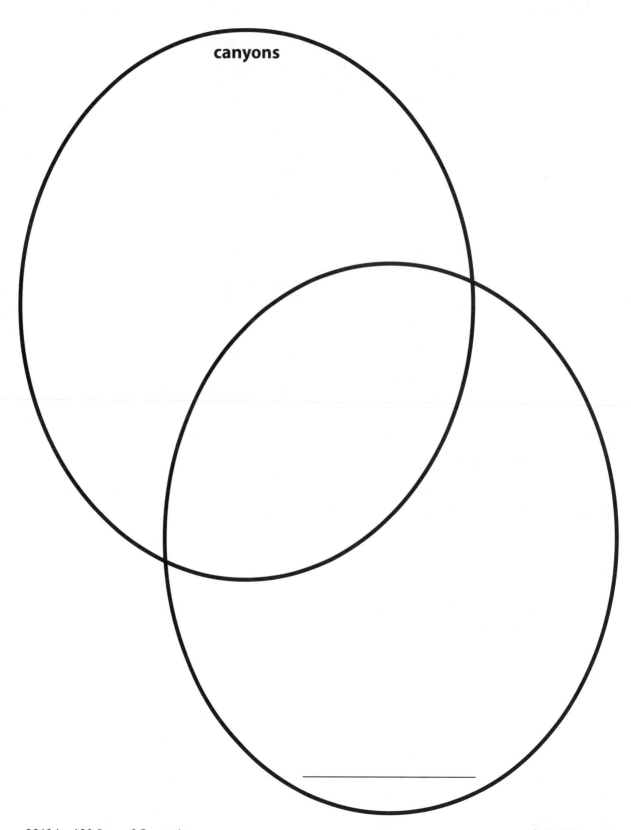

canyons

_____

Name: _____   **Date:** _____

Directions: Storms can wear down shorelines and cause flooding. People build barriers, or breakwaters, to help prevent this. Study the diagram of a breakwater, and answer the questions.

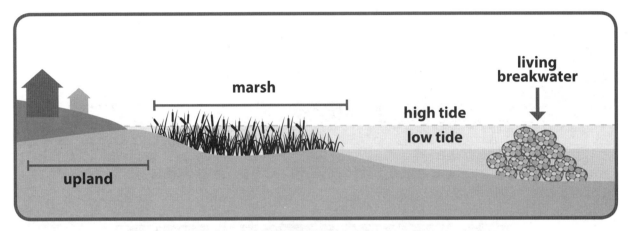

1. What happens to the marsh during high tide?

   _____

2. Why do you think homes are built on the upland area?

   _____

   _____

3. What might the term *living breakwater* mean?

   _____

   _____

4. How might this system help prevent flooding?

   _____

   _____

Creating Maps

Name: _____ **Date:** _____

Directions: Use the words in the Word Bank to label the diagram. Then, draw the diagram as it would look from up high.

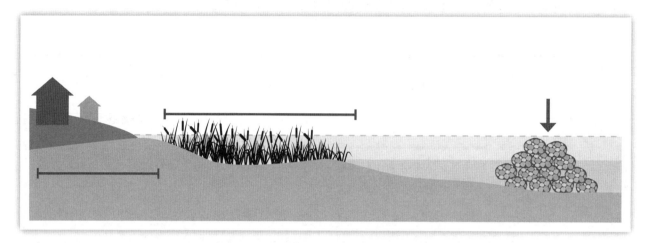

Word Bank

| marsh | upland | living breakwater |
| high tide | low tide | |

Name: _____ **Date:** _____

Directions: Read the text, and study the photo. Then, answer the questions.

Living Reef Breakwater

Part of New Jersey's coast is in trouble. Storms have worn away its shorelines. When there is no shore, beach cities can flood. Storms can bring water right into people's homes. The lack of shorelines is also a problem for wildlife. Many animals use the shore as their habitat.

The red knot bird is one of these creatures. The bird depends on New Jersey shores for its habitat. A species of red knot called the rufa has been listed as endangered. Many think this is due to its loss of habitat.

So, is there an answer? Yes! People are building breakwaters near Delaware Bay. These are barriers that help absorb the ocean's impact. They protect the shorelines and the red knot birds. These breakwaters are special. They are often made from shells. Using shells helps create habitats for sea creatures. These breakwaters become living reefs!

1. Why are New Jersey's coasts in trouble?

_____

2. Why are shorelines important?

_____

_____

3. What makes these breakwaters special?

_____

_____

Think About It

Name: _____ **Date:**_____

Directions: These photos show a beach before and after it was restored. Study the photos, and answer the questions.

1. Describe the beach before and after it was restored.

   _____

   _____

   _____

   _____

2. Do you think the restoration was successful? Why, or why not?

   _____

   _____

   _____

Name: _____ **Date:** _____

Directions: Think about how you would like to help the environment near you. Design your project step by step. Then, draw a picture of your project in the box.

Project Name: _____

Goal: _____

Step 1: First, I will _____

_____

Step 2: Then, I will _____

_____

Step 3: Finally, I will _____

_____

Geography and Me

Reading Maps

Name: _____ **Date:** _____

Directions: This map shows where goods are shipped around the word. Study the map, and answer the questions.

Global Trade

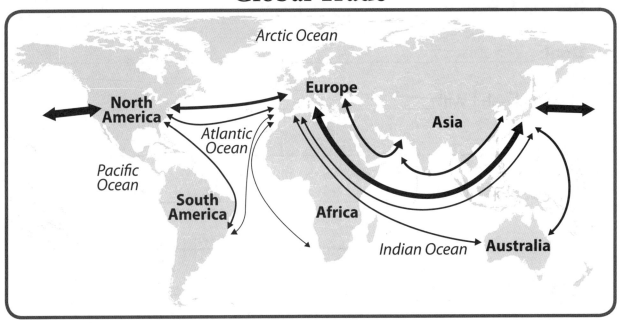

1. Where do goods that leave from the East Coast of North America go?

   _____

   _____

2. Describe the route between North America and Asia.

   _____

   _____

3. What does this map tell you about world trade?

   _____

   _____

28624—180 Days of Geography   © Shell Education

Name: _____ **Date:** _____

Directions: Follow the steps to complete the map.

Global Trade

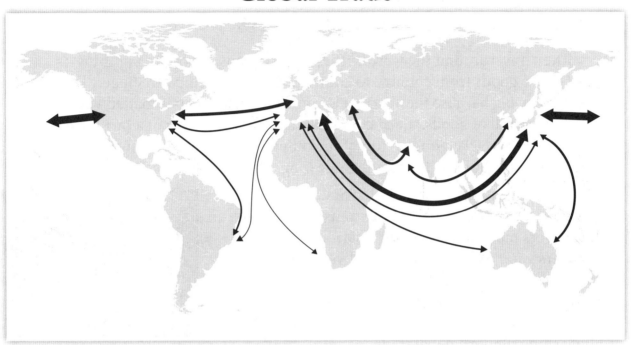

1. Label the continents. Use the Continents Word Bank to help you.

Continents Word Bank		
Africa	Asia	Europe
Australia	North America	South America

2. Label the oceans. Use the Oceans Word Bank to help you.

Oceans Word Bank			
Pacific Ocean	Indian Ocean	Arctic Ocean	Atlantic Ocean

3. Circle the regions that trade directly with North America.

Read About It

Name: _____ **Date:** _____

Directions: Read the text, and study the photo. Then, answer the questions.

Sea Monsters

They travel around the world. They cross oceans. They are big ships that move goods from country to country. Container ships are huge. They need to be big because they carry a lot of cargo. Each ship is loaded with containers full of goods. Each container is about the length of a bus. These big boxes are stacked one on top of the other. The biggest container ship in the world is about one fourth of a mile long. It's a sea monster!

The busiest port in The United States is in Los Angeles. This port is part of the transpacific route. It delivers most of its goods to Asia. The goods it delivers vary. When goods are delivered to Japan, they are usually food related. Japan has many mountains but not many places to grow food. It must import its food! This is just one example of how ships support global trade.

1. What are container ships?

_____

2. What kinds of goods does the United States deliver to Japan? Why?

_____

3. Why is the transpacific route important?

_____

_____

_____

Name: _____ **Date:** _____

Directions: This diagram shows the size of the containers found on container ships. Study the diagram. Then, answer the questions

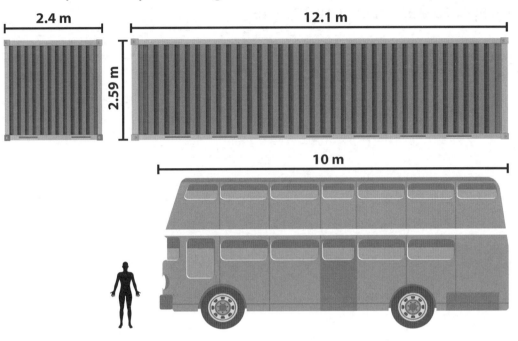

1. Is the container longer or shorter than the bus?

_____

2. Are you taller or shorter than the container? How do you know?

_____

_____

3. Do you think your classroom is longer or shorter than the container? Why do you think so?

_____

_____

4. Why do you think containers are so large?

_____

Geography and Me

Name: _____ **Date:**_____

Directions: Imagine you are piloting a container ship from North America to Asia. Which route will you take? Draw a picture, and write a paragraph to describe your journey.

_____

_____

_____

_____

_____

_____

_____

Name: _____ **Date:** _____

Directions: This map shows the path of Hurricane Katrina in 2005. Study the map closely. Then, answer the questions.

Path of Hurricane Katrina

1. Through which states did Hurricane Katrina pass?

_____

_____

2. Through which gulf did the storm pass?

_____

3. Use cardinal directions to describe the path of the storm.

_____

_____

Name: _____ **Date:** _____

Directions: Follow the steps to complete the map.

Path of Hurricane Katrina

1. The storm became a Category 1 hurricane off the east coast of Florida. Place a star there.

2. It strengthened to a Category 4 hurricane in the Gulf of Mexico. Draw a triangle there.

3. It weakened to a tropical storm in central Mississippi. Draw a square there.

4. It weakened further to a tropical depression in Tennessee. Draw a cloud there.

© Shell Education

Name: _____ **Date:** _____

Directions: Read the text, and study the photo. Then, answer the questions.

Losing Everything

Imagine losing your home, your car, your entire town. How would you feel? This happened to many people in 2005. In New Orleans, a huge storm hit. It was called Hurricane Katrina.

The hurricane began as a tropical storm. As its wind speed increased, the storm gained power. In just a few days, a hurricane formed. This storm moved along the Gulf Coast. As a result, thousands of homes and lives were lost.

The city of New Orleans suffered the most. Floods overtook the city. Streets, cars, and homes were underwater. Thousands of people were stranded. But the danger did not end once the storm passed. People had little access to food and water. When the flooding stopped, people had to work hard to rebuild their city.

1. How do you think the residents of New Orleans felt after the storm had passed?

_____

_____

2. What caused the storm to gain power?

_____

3. What happened after the storm passed New Orleans?

_____

_____

Think About It

Name: _____ **Date:** _____

Directions: This chart shows a time line of Hurricane Katrina. Study the chart closely. Then, answer the questions.

Date	Events	Wind Speed
Aug. 23	A tropical depression forms in the Bahamas.	< 39 mph
Aug. 24	It is now categorized as a tropical storm.	50 mph
Aug. 25	The storm becomes a Category 1 hurricane off the eastern coast of Florida. It passes through southern Florida.	80 mph
Aug. 26	It becomes a Category 2 hurricane in the Gulf of Mexico, off the western coast of Florida.	105 mph
Aug. 27	The storm grows stronger. It becomes a Category 3 hurricane in the Gulf of Mexico.	115 mph
Aug. 29	The storm makes landfall in Louisiana as a Category 4 hurricane.	145 mph
Aug. 29	It weakens to a tropical storm in central Mississippi.	60 mph
Aug. 30	It weakens to a tropical depression in Tennessee.	< 39 mph

1. When was Hurricane Katrina the strongest? What happened?

_____

_____

2. Hurricanes are rated on a scale of 1 through 5. Which number represents the strongest storm? How do you know?

_____

_____

3. Which state do you think received the most damage: Florida, Louisiana, Mississippi, or Tennessee? Why do you think so?

_____

_____

© Shell Education

Name: _____ **Date:** _____

Directions: Think about the strongest storm you have ever seen. Write what happened during that storm. Then, write about what happened during Hurricane Katrina.

The Storm I Saw	Hurricane Katrina

Reading Maps

Name: _____ **Date:** _____

Directions: Study the map closely. Then, answer the questions.

U.S. Capitol

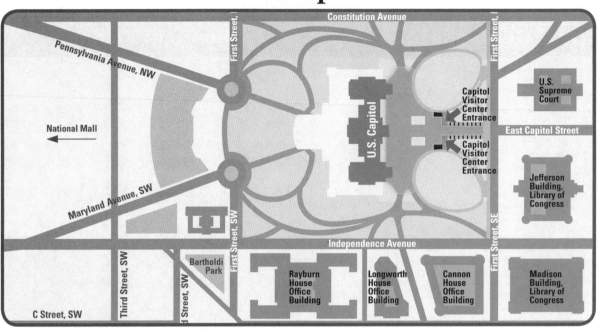

1. Which building is the focus of the map? How do you know?

   _____

   _____

2. List at least two buildings that surround the U.S. Capitol building.

   _____

   _____

3. How do some of the street names reflect the history and purpose of the Capitol?

   _____

   _____

Name: _____ **Date:**_____

Directions: Follow the steps and answer the questions.

U.S. Capitol

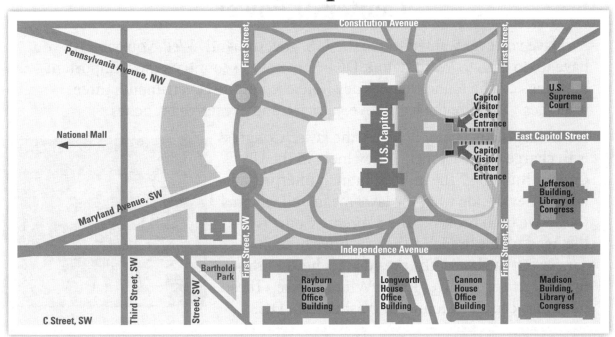

1. Draw two routes from the Supreme Court to the Rayburn House Office Building. Which is shorter?

   _____

2. Draw a route from Pennsylvania Avenue to the Library of Congress Jefferson building. List the directions. Include which ways to turn and the street names.

   _____

   _____

   _____

   _____

   _____

Creating Maps

Read About It

Name: _____ **Date:** _____

Directions: Read the text, and study the photo. Then, answer the questions.

Center of Power

Have you ever wanted to see the U.S. government? Well, you can. All you have to do is go to Washington, DC! Some of the country's most important buildings can be found there. They hold each of the government's three branches. They are the legislative, judicial, and executive branches.

The legislative branch makes the laws. Congress is in charge of this branch. They meet in the U.S. Capitol Building. The judicial branch interprets the laws. They decide what the laws mean and whether they have been broken. The Supreme Court is the head of this branch. The executive branch carries out the laws. The president is head of this branch. He or she lives in the White House. These buildings can all be found in Washington, DC!

1. What does the legislative branch do?

_____

_____

2. What does the judicial branch do?

_____

_____

3. Why do you think the U.S. government buildings are in the same area?

_____

_____

Name: _____ **Date:**_____

Directions: This chart shows the three branches of the U.S. government. Study the chart, and answer the questions.

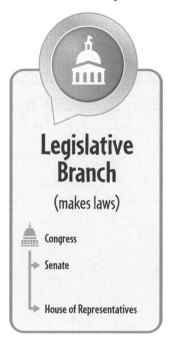

Legislative Branch
(makes laws)

Congress
→ Senate
→ House of Representatives

Executive Branch
(carries out laws)

President
→ Vice President
→ Cabinet

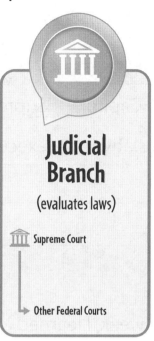

Judicial Branch
(evaluates laws)

Supreme Court
→ Other Federal Courts

1. What makes up the judicial branch?

_____

_____

2. Who is part of the executive branch?

_____

_____

3. Which branch is in charge of making laws?

_____

Name: _____ **Date:** _____

Geography and Me

Directions: Write a letter to a friend telling what you learned about the U.S. Capitol and the three branches of government. Use the Word Bank to help you complete your sentences.

Word Bank			
U.S. Capitol	Supreme Court	White House	government
legislative	executive	judicial	

Dear _____,

This week, I learned about _____

_____

_____

In Washington, DC, there are _____

_____

_____

The three branches of government are _____

_____

_____

Your friend,

_____

© Shell Education

Name: _____   **Date:** _____

Directions: The Yukon is a territory in Canada. Study the map carefully. Then, answer the questions.

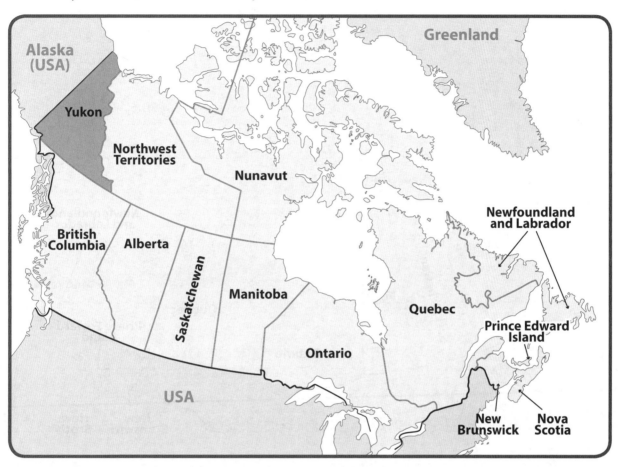

1. Which three regions border the Yukon?

    _____

2. Judging by its location, what do you think the climate of the Yukon is like in the winter?

    _____

3. Do you think many people live in the Yukon? Why, or why not?

    _____

    _____

Name: _____ Date:_____

Directions: Use the Word Bank to label the provinces of northwestern Canada. Shade the Yukon Territory.

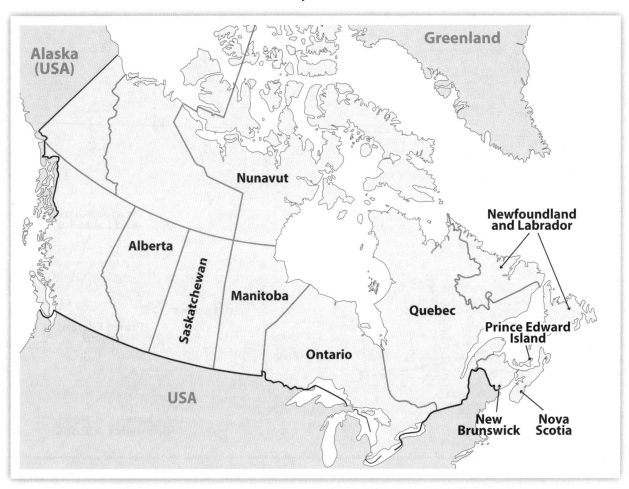

Word Bank		
Yukon	British Columbia	Northwest Territories

Challenge: Canada is bordered by three oceans. Label them on the map. Then, ask an adult to help you check your answers.

Name: _____ **Date:** _____

Directions: Read the text, and study the photo. Then, answer the questions.

The Yukon Life

Life can be cold, even freezing, in the Yukon. The cold season can feel long and brutal. But it's a different story in the summer. The season is short, but each summer day is long. Because of its northern position, the Yukon experiences many hours of sunlight during the summer.

The Yukon does not make money from farming. Much of its soil is permanently frozen. In the past, people in the Yukon relied on mining as a resource. Copper, lead, and silver have all been mined from the land. But today, the region makes much of its money through tourism.

Even though the Yukon can be very cold, people love to visit. There are many things to do in the Yukon. During the summer, people canoe, hike, and fish. When it is cold, tourists can ski, snowboard, and sled. Some people even dog sled!

1. How does the Yukon's climate affect farming?

_____

_____

2. How does the Yukon make money?

_____

3. What do people like to do in the Yukon?

_____

_____

Think About It

Name: _____ **Date:** _____

Directions: These photos show the Yukon Territory in the summer and winter. Study the photos, and answer the questions.

Yukon in summer   **Yukon in winter**

1. Describe the Yukon in the summer.

_____

_____

2. Describe the Yukon in the winter.

_____

_____

3. How do the seasons affect what you can do for fun in the Yukon?

_____

_____

_____

Name: _____ **Date:** _____

Directions: Think about your region. What activities are available close by? What activities are only available far away? Use your ideas to complete the chart.

Activities Near Me	Activities Far Away

Geography and Me

Reading Maps

Name: _____ Date:_____

Directions: This is a map of the Los Angeles Aqueduct. Aqueducts transport water from one area to another. Study the map, and answer the questions.

1. The Los Angeles Aqueduct is east of which mountain range?

   _____

   _____

2. The runs along what river?

   _____

   _____

3. Why do you think the aqueduct was built?

   _____

   _____

   _____

   _____

   _____

   _____

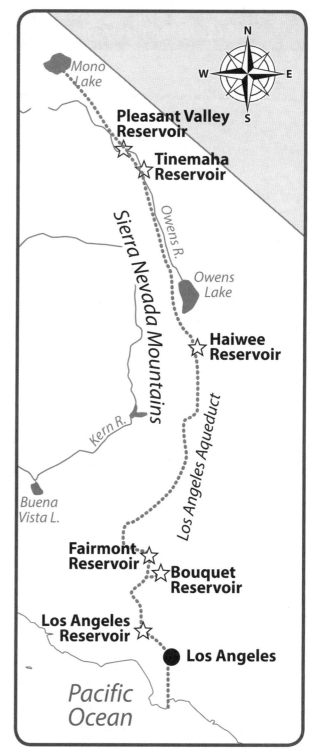

Name: _____ **Date:** _____

Directions: Follow the steps to complete the map.

1. Add a title to the map.

2. Add a compass rose to the map.

3. Label the Owens River.

4. Trace in blue other rivers and lakes shown.

5. Trace the route the water would take from Mono Lake to Los Angeles.

6. Describe the route your traced.

    _____

    _____

    _____

    _____

    _____

7. Circle where the water empties into the ocean.

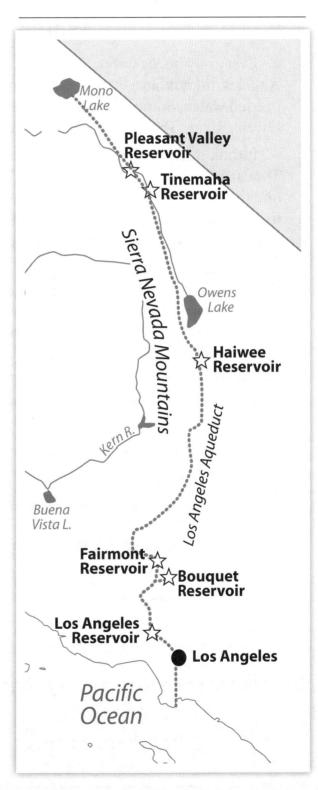

_____

Mono Lake

Pleasant Valley Reservoir

Tinemaha Reservoir

Sierra Nevada Mountains

Owens Lake

Haiwee Reservoir

Los Angeles Aqueduct

Kern R.

Buena Vista L.

Fairmont Reservoir

Bouquet Reservoir

Los Angeles Reservoir

Los Angeles

Pacific Ocean

Creating Maps

Read About It

Name: _____ **Date:** _____

Directions: Read the text, and study the photo. Then, answer the questions.

Water Wanted!

Every city needs water. It is a key resource. There was a time when Los Angeles did not have water. In the early 1900s, there was a drought. The city needed water. So people looked for a new source. They found water in the Owens River. This was the start of the Los Angeles Aqueduct.

It took many years to complete the project. Thousands of workers joined forces. They used one million barrels of cement. They built close to 300 miles of pipeline. The project took blood, sweat, and tears. But, in 1913, it was completed.

Once it was done, many moved to the city. At the time, there were only a few hundred thousand people living in the area. But that changed with the aqueduct! There was now enough water to support millions. Today, almost four million people live in the city.

1. Why was the aqueduct built?

   _____

2. Describe the building of the Los Angeles Aqueduct.

   _____

   _____

3. What happened to the city after the aqueduct was completed?

   _____

   _____

Name: _____ **Date:** _____

Directions: This bar graph shows how much freshwater California used from 1985 and 2010. Study the graph closely. Then, answer the questions.

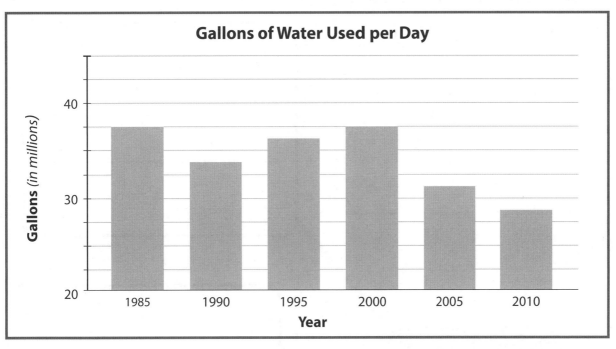

1. During which year did California use the least water?

   _____

2. During which years do you think California may have suffered from droughts? Why?

   _____

   _____

3. What can people do to conserve, or save, water in their daily lives?

   _____

   _____

   _____

© Shell Education

Think About It

Name: _____ **Date:** _____

Directions: Think about ways you can conserve water. Write five ways on the graph.

How I Can Conserve Water

Name: _____   **Date:** _____

Directions: Mexico City is one of the largest cities in the world. Study its location. Then, answer the questions.

1. Describe Mexico City's location in Mexico.

   _____

2. List three countries that border Mexico.

   _____

   _____

   _____

3. What are the three closest bodies of water to Mexico City?

   _____

   _____

   _____

Creating Maps

Name: _____ **Date:**_____

Directions: Follow the steps to complete the map.

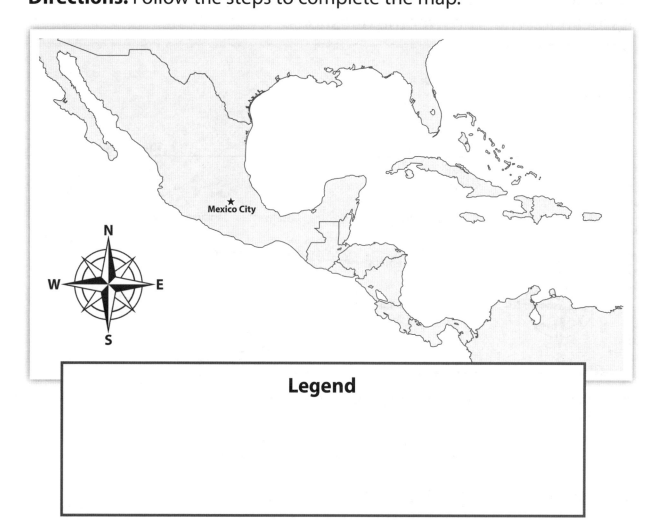

Mexico City

N
W E
S

Legend

1. Label the bodies of water closest to Mexico City.

2. Label the countries that border Mexico.

3. Choose a color, and use it to shade Mexico.

4. Choose a different color to shade the countries that border Mexico.

5. Create a legend to show what your colors mean.

Name: _____ **Date:**_____

Directions: Read the text, and study the photo. Then, answer the questions.

A Vibrant Culture

Mexico City is a center of arts and culture. There are many museums to visit. There are also sports centers. Sports fans travel in crowds to watch their favorite teams. For these reasons and others, tourists flock to the city. They know that it has much to offer.

Mexico City is the largest city in the country. It was built on top of an ancient Aztec city. The Aztecs lived in central Mexico for hundreds of years. Then, the Spanish invaded. They took the land from the Aztecs. They destroyed Aztec temples. Then, the Spanish built a new city. Aztec culture is still a part of Mexico City. It is home to Aztec ruins.

1. How does the photo show the culture of Mexico City?

_____

_____

2. What things can people do in Mexico City?

_____

_____

3. Who were the Aztecs?

_____

_____

Think About It

Name: _____ **Date:** _____

Directions: The Aztecs lived in Mexico long ago. These are some of the few Aztec buildings left. Study the photo, and read the caption. Then, answer the questions.

Aztec ruins in Mexico City

1. Describe the ruins.

_____

_____

2. Why might someone want to visit this structure?

_____

_____

3. How does this show part of Mexico City's culture?

_____

_____

28624—*180 Days of Geography*   © *Shell Education*

Name: _____ **Date:** _____

Directions: Compare where you live to Mexico City. How are the two places similar and different?

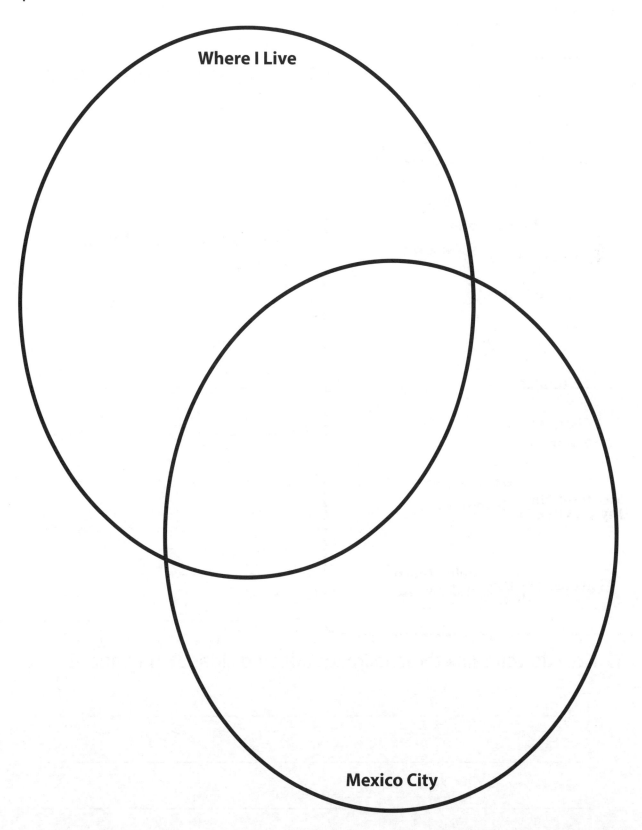

Where I Live

Mexico City

Geography and Me

Reading Maps

Name: _____ **Date:** _____

Directions: The Florida East Coast Railway was created in the 1890s. It helped connect Florida with the rest of the world. Study the map, and answer the questions.

Legend

☐ railroad station

☒ railroad

▨ Florida East Coast Railway

Boston
Providence
New York
Philadelphia
Baltimore
Washington, DC
Richmond
Florence
Charleston
Savannah
Jacksonville
St. Augustine
Ormond
Daytona Beach
Palm Beach
Fort Lauderdale
Hollywood
Miami

1. What symbol shows railroad stations?

   _____

   _____

2. How do you think these railroads affected Florida's trade?

   _____

   _____

   _____

   _____

   _____

   _____

3. How do you think these railroads affected Florida's migration?

   _____

   _____

   _____

Name: _____ **Date:** _____

Directions: Follow the steps to complete the map.

_____

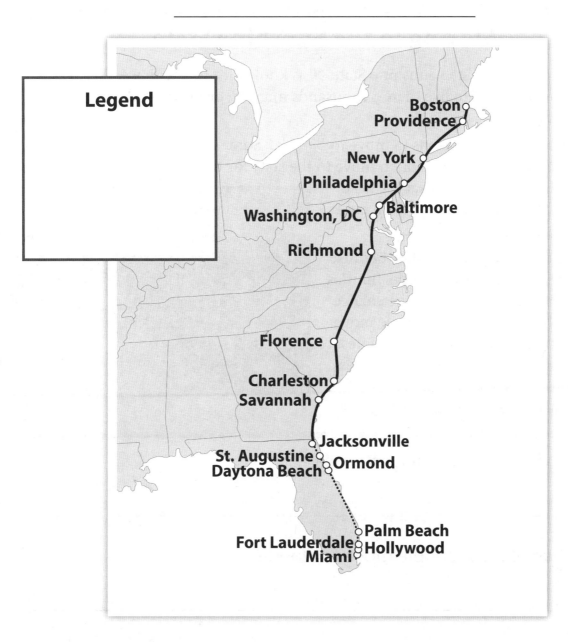

1. Add a title to the map.

2. Add a compass rose next to the map.

3. Create a legend that shows symbols for railroad stops and railroad lines.

Read About It

Name: _____ **Date:** _____

Directions: Read the text, and study the photo. Then, answer the questions.

Moving to the Sunshine State

Florida is called the Sunshine State. It is known for its warm weather. Many people want to live there. Thousands migrate, or move, to the state each year!

There are many reasons to move to Florida. Some people are drawn by its warm climate. Seniors often move to the state for this very reason. After they retire from their jobs, they head south. These seniors move from cold climates to a sunny paradise.

Others migrate to the state because they are looking for new opportunities. They want a better life and the chance to find better jobs. Such people move from other parts of the United States. They also move from other countries. For these reasons, the population of the Sunshine State continues to grow.

1. What does the word *migrate* mean?

_____

2. List two reasons people move to Florida.

_____

_____

3. Would you like to live in Florida? Why, or why not?

_____

_____

Name: _____ **Date:**_____

Directions: This chart shows the 10 most popular theme parks in 2015. Study the chart, and answer the questions.

World Rank	Theme Park	Location
1	The Magic Kingdom at Walt Disney World	Orlando, Florida
2	Disneyland	Anaheim, California
3	Tokyo Disneyland	Tokyo, Japan
4	Universal Studios Japan	Osaka, Japan
5	Tokyo DisneySea	Tokyo, Japan
6	Disney's Epcot	Orlando, Florida
7	Disney's Animal Kingdom	Orlando, Florida
8	Disney's Hollywood Studios	Orlando, Florida
9	Disneyland Paris	Paris, France
10	Universal Studios	Orlando, Florida

1. How many of the top 10 theme parks in the world are in Florida?

_____

2. How do you think these theme parks affect Florida's economy?

_____

_____

3. How do you think theme parks encourage tourism?

_____

_____

Geography and Me

Name: _____ **Date:** _____

Directions: Write and draw to show three things you learned about Florida.

_____

_____

_____

_____

_____

_____

ANSWER KEY

There are many open-ended pages and writing prompts in this book. For those activities, the answers will vary. Answers are only given in this answer key if they are specific.

Week 1 Day 1 (page 15)
1. Moloka'i
2. O'ahu
3. Hawai'i
4. Kaua'i

Week 1 Day 2 (page 16)
1. Starting at the top and moving clockwise: North, Northeast, East, Southeast, South, Southwest, West, Northwest
2. Nashville
3. Chattanooga
4. west

Week 1 Day 3 (page 17)
1. The lightest shade in the legend should be circled.
2. Answers may include that the legend uses labels to show what the shades mean.
3. Answers may include that cities are darkest because more people live there.

Week 1 Day 4 (page 18)
1. California
2. New York
3. New York, Florida, Illinois, Texas, California

Week 2 Day 1 (page 20)
1. equator
2. longitude
3. latitude

Week 2 Day 3 (page 22)
1. lake
2. D5
3. C2
4. library

Week 2 Day 4 (page 23)
1. Atlanta
2. At least three of the following should be listed: Atlanta, Statesboro, Savannah, Athens, Valdosta, Columbus, Macon, Albany, and Augusta.
3. Columbus

Week 2 Day 5 (page 24)
1. Topeka
2. C3
3. A2

Week 3 Day 1 (page 25)
1. Answers may include that a ridge is a peak or a mountain.
2. Answers may include that a valley is a low area between mountains.
3. Answers may include that the pattern is ridge, valley, ridge.

Week 3 Day 3 (page 27)
1. water
2. Sandstone is harder to wear down than limestone.
3. The valleys might have been eroded over time.

Week 3 Day 4 (page 28)
1. the Ridge and Valley Province and the Piedmont Province
2. Georgia, Maryland, Pennsylvania, and Virginia should be outlined.
3. New England-Acadian Province

Week 4 Day 1 (page 30)
1. California, Nevada, Utah, and Arizona
2. California
3. Lancaster, Barstow, and Victorville

Week 4 Day 2 (page 31)
1. Colorado River
2. Yes, California has mountains. There are triangle symbols that stand for mountains in the map of California.
3. Phoenix

Week 4 Day 3 (page 32)
1. Answers may include that days are scorching hot, nights are freezing, and the weather can change from day to day.
2. Answers may include that many plants can't survive the changing desert weather.
3. An ecosystem is a community of living things.

ANSWER KEY (cont.)

Week 4 Day 4 (page 33)

1. July
2. January and December
3. Answers should explain in what month students would visit the Mojave Desert based on average temperatures.

Week 5 Day 1 (page 35)

1. C1
2. C4
3. Answers should include any five of the following states: Minnesota, Wisconsin, Iowa, Missouri, Illinois, Kentucky, Tennessee, Arkansas, Mississippi, and Louisiana.

Week 5 Day 3 (page 37)

1. A natural border is one that has been shaped by nature, not humans.
2. Trade brings jobs and money to people living in the region.
3. Answers should include any three of the following: coal, oil, grain, and steel.

Week 5 Day 4 (page 38)

1. Greenville and Rosedale
2. The border closely follows the river, but it does not match exactly.
3. Vicksburg; it is closest to the city.

Week 6 Day 1 (page 40)

1. walking, biking, and driving
2. walking or biking
3. walking

Week 6 Day 3 (page 42)

1. A student might see it as a place to learn about plants and animals.
2. An athlete might see it as place to play different sports.
3. Answers may include that people see the park differently because they have different interests.

Week 6 Day 4 (page 43)

1. 25,000 trees
2. 25,000 trees
3. 68,000 trees
4. Example: *I think they will meet their goal because they were already well on their way by 2016.*

Week 7 Day 1 (page 45)

1. Golden Gate Bridge and San Francisco-Oakland Bay Bridge
2. Golden Gate Bridge
3. Answers may include Alcatraz Island, Coit Tower, SF MOMA, Candlestick Point, Lake Merced, and Golden Gate Park.

Week 7 Day 3 (page 47)

1. eat at restaurants, see a movie, and go to museums
2. Answers may include that there are many things to do and it is convenient.
3. Answers may include that cities are noisy and crowded.

Week 7 Day 4 (page 48)

1. Four subway lines pass through San Francisco.
2. Three subway lines pass through MacArthur.
3. Answers may include that more people live in San Francisco or that more people want to visit the city.

Week 8 Day 1 (page 50)

1. Utah, Colorado, Arizona, and New Mexico
2. Colorado Plateau area
3. Answers may include that the area has rivers and lakes or that it is home to the Grand Canyon.

Week 8 Day 3 (page 52)

1. Mesas become buttes through erosion.
2. Answers may include that erosion involves wearing down or eating away at rocks.
3. Together, the large butte and the thin butte look like a mitten.

Week 8 Day 4 (page 53)

1. Answers may include that the tops of the landforms are all flat.
2. A mesa is wider than a butte.
3. Answers may include that a plateau is a wide landform with a flat top.

Week 9 Day 1 (page 55)

1. It gets more sunlight in June because it is tilted toward the sun.
2. It gets more sunlight in December because it is tilted toward the sun.
3. December is a warm month because the Southern Hemisphere gets more sunlight.

ANSWER KEY (cont.)

Week 9 Day 2 (page 56)

spring in September; summer in December; fall in March; winter in June

Week 9 Day 3 (page 57)

1. Answers may include that *revolves* means "goes around."
2. It is winter in the Northern Hemisphere, and the weather is cooler.
3. Answers may include that it is winter in the Northern Hemisphere when it is summer in the Southern Hemisphere; their seasons are opposite.

Week 9 Day 4 (page 58)

1. Southern Hemisphere; it is colder in June.
2. Northern Hemisphere; it is colder in December.
3. decrease; it would be the middle of winter.

Week 10 Day 1 (page 60)

1. mining
2. fishing
3. mining, fishing, dairy, and poultry

Week 10 Day 3 (page 62)

1. milk, yogurt, and cheese
2. The demand for dairy products continues to grow.
3. It creates gases that can lead to global warming.

Week 10 Day 4 (page 63)

1. soybeans
2. wheat
3. soybeans
4. Example: *I would grow wheat because it has the most bushels per acre.*

Week 11 Day 1 (page 65)

1. Answers may include that gold miners faced animals and cold conditions.
2. Answers may include that gold miners had to find ships and travel for many months.
3. Answers will vary but should include reasons why they chose one of the routes.

Week 11 Day 2 (page 66)

Students may use any three colors or patterns, but they should show roughly the routes below.

Week 11 Day 3 (page 67)

1. Answers may include that people didn't believe the gold was real and later people caught gold fever.
2. Answers may include that miners lived rough lives, there was no one to govern them, life at mining camps could be violent, and mining was hard work.
3. pans, picks, and cradles

Week 11 Day 4 (page 68)

1. the riffles
2. Answers may include that it is larger.
3. Answers may include that the faster they work, the better the chances are that they'll find gold.

Week 12 Day 1 (page 70)

1. Arctic Ocean
2. Answers should include that it is very far north.
3. Answers may include that the climate is very cold and icy.

ANSWER KEY (cont.)

Week 12 Day 3 (page 72)

1. Answers may include that polar bears have a layer of fat that helps keep them warm, they have uneven skin on their feet to prevent slipping, and they have sharp claws that help them catch their prey.
2. Answers may include that you would not find a polar bear near warm places because they need ice floes to survive.
3. Answers may include that its fur helps it blend in with its surroundings.

Week 12 Day 4 (page 73)

1. 1978
2. Sea ice has decreased in the last 30 years.
3. Answers should include that this trend means there is less sea ice for polar bears to live on.

Week 13 Day 1 (page 75)

1. Canada's population is far lower than that of the United States.
2. Answers should include three of the following: Seattle, San Francisco, Los Angeles, New Orleans, Miami, Minneapolis-St. Paul, Chicago, Detroit, Toronto, Boston, New York, Philadelphia, and Mexico City.
3. More people live near the coast. Coastal areas have high populations.

Week 13 Day 3 (page 77)

1. Answers may include that cities grow around ports, and ports are centers of jobs and trade.
2. They bring more businesses and attractions, such as theaters and museums, and cultures become more dynamic and diverse.
3. Answers may include that people like the activities, nature, or that they want to live in a diverse community.

Week 13 Day 4 (page 78)

1. Chicago, IL
2. Its population will increase.
3. New York's population will likely increase since it increased from 2000 to 2010.

Week 14 Day 1 (page 80)

1. Any three of the following states should be listed: Nevada, Oregon, Idaho, California, Utah, Wyoming, Colorado, and Arizona
2. the Rocky Mountains and the Sierra Nevada Mountains
3. Answers may include the Mojave Desert and the Colorado Plateau.

Week 14 Day 3 (page 82)

1. The word *basin* means bowl.
2. The climate is dry, and temperatures change from hot to cold throughout the day.
3. Mountain ranges block winds from bringing clouds.

Week 14 Day 4 (page 83)

1. Shoshone
2. Shoshone and Ute
3. Washoe, Mono, and Paiute

Week 15 Day 1 (page 85)

1. magma
2. Answers may include that eruptions from the man vent are much higher and include gas and ash, while eruptions from the side vent create lava flows.
3. gas and ash

Week 15 Day 2 (page 86)

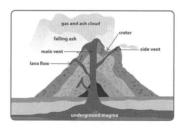

Week 15 Day 3 (page 87)

1. Answers may include that the volcano in the photograph is erupting, like Kilauea.
2. Volcanoes made the Hawaiian Islands. For millions of years, volcanoes erupted. The lava cooled to become land.
3. It is the largest shield volcano in the world.

ANSWER KEY *(cont.)*

Week 15 Day 4 (page 88)

1. Answers may include that a shield volcano is wide, has a gentle slope, and is shaped like a large dome.
2. Answers may include that the stratovolcano is more cone-like, while the shield volcano is shaped like a dome.
3. Answers may include that caldera volcano is wider and has severed cones.

Week 16 Day 1 (page 90)

1. state government
2. federal government
3. county government; answers should reference the legend.

Week 16 Day 3 (page 92)

1. to change
2. federal, state, and cities or counties
3. The 10th Amendment says that the state has any powers that the federal government does not.

Week 16 Day 4 (page 93)

1. Answers may include declaring war, coining money, and overseeing trade. These issues affect the whole country.
2. Answers may include maintaining parks, maintaining police department, maintaining fire department. These issues only affect the local community.

Week 17 Day 1 (page 95)

1. the East
2. the West
3. Answers may include North Dakota, South Dakota, Nebraska, Iowa, Kansas, and Illinois.

Week 17 Day 3 (page 97)

1. They are often used for timber.
2. private land
3. in the West
4. Answers may include that public forests are important because they preserve natural land, and private forests are important because they help people make money.

Week 17 Day 4 (page 98)

1. Answers should describe that it is a small part of the world population.
2. The amount of timber the United States uses is larger.
3. Answers should include that the United States' timber use is higher than its population.

Week 18 Day 1 (page 100)

1. Washington, Idaho, Nevada, and California
2. Pacific Ocean
3. Answers may include fishing or tourism.

Week 18 Day 2 (page 101)

Starting at the top and moving clockwise: Washington, Idaho, Nevada, California, and Pacific Ocean.

Week 18 Day 3 (page 102)

1. Answers may include that nature creates jobs for fishers, farmers, and other people who make their living from the land.
2. timber and fish
3. Answers should explain that there are only so many of these resources, so if they are overused, they might not recover.

Week 18 Day 4 (page 103)

1. 2004
2. Answers should describe that there were more jobs than in 2014 because the data are trending up.
3. Answers should include reasons to support opinions.

Week 19 Day 1 (page 105)

1. Tennessee
2. Six states; Tennessee, Kentucky, Illinois, Missouri, Arkansas, and Oklahoma.
3. The water route is longer. Explanations may include that this route has more turns, while the northern route is more direct.

Week 19 Day 3 (page 107)

1. The United States wanted the land's gold and other resources.
2. President Jackson, Congress, and other white people who wanted the Cherokee land.
3. Answers may include that when people choose to move, they have more freedom.

ANSWER KEY *(cont.)*

Week 19 Day 4 (page 108)

1. Answers may include that when gold was found, white people wanted the land and its resources. Congress passed the Indian Removal Act so that they could have access to the land.
2. They brought their case to court.
3. 1838

Week 20 Day 1 (page 110)

1. Three Quebec cities should be listed.
2. Answers may include that they are close to water or that they are mostly in the southern part of the province.
3. Near the water because that is where more cities are located.

Week 20 Day 2 (page 111)

The Yukon, Northwest Territories, and Nunavut should be one color. A box should be drawn around British Columbia, the Yukon, and the Northwest territories. The rest of the provinces should be a shaded a different color.

Week 20 Day 3 (page 112)

1. Quebec was founded by the French. Many people who live in Quebec have French ancestors.
2. People in Quebec speak French, are taught in French, and watch French TV shows.
3. Answers may include that language affects what people eat, what they wear, and what they watch on TV.

Week 20 Day 4 (page 113)

1. English and French
2. Any three of the following should be listed: Punjabi, Italian, Spanish, German, Cantonese, Tagalog, and Arabic.
3. Answers may include that many languages are spoken in Canada. This shows that different cultures can share the same country.

Week 21 Day 1 (page 115)

1. Gulf of Mexico, Atlantic Ocean, and Lake Okeechobee
2. Lake Okeechobee
3. Answers may describe a marsh as muddy and watery.

Week 21 Day 2 (page 116)

The Everglades is the shaded region. The Gulf of Mexico is on the left. The Atlantic Ocean is on the right. Lake Okeechobee is in the middle of the state.

Week 21 Day 3 (page 117)

1. A wetland is a region that is covered in water, such as marshes and swamps.
2. Answers may include the sawgrass, turtles, snakes, bobcats, otters, birds, crocodiles, and manatees.
3. People were draining the marsh in order to create farmland.

Week 21 Day 4 (page 118)

1. June is the rainiest; December is the driest.
2. Answers should include the summer months.
3. Answers may include that animals need to adapt to live in wet, rainy conditions in the summer and dry conditions in the winter.

Week 22 Day 1 (page 120)

1. tundra
2. grassland, desert, mixed forest
3. taiga

Week 22 Day 3 (page 122)

1. Answers may include that the taiga has long, cold winters and short, cool summers.
2. Answers may include that they absorb as much sunlight as possible and their branches are angled toward the ground.
3. Answers may include wolves, squirrels, lynxes, moose, and reindeer.

Week 22 Day 4 (page 123)

1. Answers should be close to 1865 and 1885.
2. Answers should be close to 1865 and 1885.
3. Answers may include that as the population of the hare goes up, so does the population of the lynx.

Week 23 Day 1 (page 125)

1. Native peoples
2. the French, English, and Spanish

ANSWER KEY *(cont.)*

Week 23 Day 3 (page 127)

1. The Louisiana Purchase was land that the United States bought in 1803.
2. France or Napoleon
3. Answers may include that $15 million is not a lot of money for the amount of land the United States purchased.

Week 23 Day 4 (page 128)

1. Answers should explain that there are fewer states in the 1854 map.
2. Answers should include that the Nebraska territory has become several different states. It including parts of what is now North Dakota, South Dakota, Montana, Wyoming, Colorado, and Nebraska.
3. Answers may include that the territories became states.

Week 23 Day 5 (page 129)

Answers may include that the colonies were still under British control in 1750, the French and Spanish still controlled parts of what would become the United States in 1750, and today, the United States is an independent country with 50 states.

Week 24 Day 1 (page 130)

1. Wyoming
2. 21 miles
3. 30 miles

Week 24 Day 3 (page 132)

1. Answers may include that Old Faithful erupts regularly.
2. Answers may include seeing Old Faithful, visiting the hot springs, soaking in the Boiling River, taking a boat tour, or visiting the museum.
3. the museum

Week 24 Day 4 (page 133)

1. Answers may include that a person can see across the lake and that a person can also see nearby fish.
2. Answers may include that a person can get a close-up look at all of the lake's features.
3. Answers may include that their experiences will be different depending on what they do and where they are.

Week 25 Day 1 (page 135)

1. Dominican Republic
2. Caribbean Sea, Gulf of Mexico, and Atlantic Ocean
3. North America and South America

Week 25 Day 3 (page 137)

1. Answers may include its warm weather, clear waters, and its plants and animals.
2. sugar, bananas, and coffee
3. Answers may include that exports make money, while imports cost money.

Week 25 Day 4 (page 138)

1. 2016
2. 2014
3. Import answers may include that the bar for imports is higher each year.
4. spent more money

Week 26 Day 1 (page 140)

1. Wyoming, Louisiana, West Virginia, and Utah.
2. Nevada, Vermont, and Maryland.

Week 26 Day 3 (page 142)

1. Answers may include that women in the early 1900s were told to stay at home, they were not allowed to vote, and working was looked down upon.
2. Answers may include that women entered the workforce to fill the jobs the men had left.
3. Answers may include that women have gained the right to vote and that many women make up almost half of the U.S. workforce.

Week 26 Day 4 (page 143)

1. Answers should be close to $13,000.
2. Answers should be close to $11,000.
3. It has decreased; answers should include that the lines are closer together on the right side of the graph.

Week 27 Day 1 (page 145)

1. Charlestown, Medford, Malden, Lexington, and Menotomy
2. twice
3. Their routes met around Menotomy.

ANSWER KEY *(cont.)*

Week 27 Day 3 (page 147)

1. Paul Revere was a Patriot who rode to warn people.
2. He rode to warn people about the British. He told them to get ready to fight.
3. Example: *Yes, because of Paul Revere, the Patriots were ready for the British.*

Week 27 Day 4 (page 148)

1. near Lexington and Concord
2. near Concord
3. Answers may include that the British may have won the battle instead.

Week 28 Day 1 (page 150)

1. The tops of the landforms are all flat and have a resistant rock layer.
2. A plateau is much larger than a mesa.
3. Answers may include that the canyon has steep sides or looks like a steep valley.

Week 28 Day 2 (page 151)

The large land form in the far background should be labeled as a plateau. The wider landforms in the foreground are mesas, and the thinner ones are buttes.

Week 28 Day 3 (page 152)

1. The river rushed over the rocks and broke them down. Over time, steep walls formed as the river sunk lower.
2. The river has broken down the rock above. The river continues to carve away at the canyon.
3. The top of the canyon should be circled.

Week 28 Day 4 (page 153)

1. These years should have green circles: 1935, 1955, 1965, 1975, 1985, 1995, and 2015.
2. 1945 and 2005
3. Answers should include details from the chart.

Week 29 Day 1 (page 155)

1. It is flooded.
2. Answers may include that the upland area is higher and would not flood regularly.
3. Answers may include that it is made of living things.
4. Answers may include that the marsh will flood before the homes flood.

Week 29 Day 3 (page 157)

1. Storms have worn away the shorelines.
2. They protect cities from flooding and are home to wildlife.
3. They provide habitats for sea creatures.

Week 29 Day 4 (page 158)

1. Answers should explain that the beach is wider after it was restored.
2. Answers should be supported by details from the photos.

Week 30 Day 1 (page 160)

1. Europe and South America
2. Answers should explain that the route goes across the Pacific Ocean.
3. Answers may include that all parts of the world are connected through trade.

Week 30 Day 3 (page 162)

1. Container ships are large cargo ships that carry goods between countries.
2. food
3. Answers may include that many goods are sent between the United States and Asia along that route.

Week 30 Day 4 (page 163)

1. longer
2. Shorter; answers may include the height markings or the illustration of the person.
3. Answers should be supported by details from the diagram.
4. Answers may include that they ship large amounts of goods.

Week 31 Day 1 (page 165)

1. Florida, Louisiana, Mississippi, and Tennessee
2. Gulf of Mexico
3. Answers should describe how the storm started in the Bahamas and moved west through Florida. Then, the storm moved north through the Gulf of Mexico and the southern United States.

ANSWER KEY *(cont.)*

Week 31 Day 3 (page 167)

1. Answers may include that they were likely sad and scared.
2. increased wind speed
3. Answers may include that people struggled because they did not have enough food or water, or that they worked hard to rebuild the city.

Week 31 Day 4 (page 168)

1. August 29; It made landfall in Louisiana.
2. 5; answers may include that in the chart, the larger numbers have higher wind speeds.
3. Louisiana; answers should explain that the winds were the strongest or the storm was still strengthening or weakening in the other states.

Week 32 Day 1 (page 170)

1. The U.S. Capitol is the focus of the map. It is in the middle of the map.
2. Answers should include two of the following: U.S. Supreme Court, Library of Congress Jefferson Building, Library of Congress Madison Building, Cannon House Office Building, Longworth House Office Building, or Rayburn House Office Building.
3. Answers may include Constitution Ave., Independence Ave., or different state names.

Week 32 Day 3 (page 172)

1. The legislative branch makes the laws.
2. The judicial branch interprets the laws.
3. Answers may include that being in the same area helps them communicate with each other.

Week 32 Day 4 (page 173)

1. the Supremem Court and other in federal courts
2. the president, vice president, and the cabinet
3. the Legislative Branch

Week 33 Day 1 (page 175)

1. Northwest Territories, British Columbia, and Alaska
2. Answers may include that the climate is very cold in the winter.
3. Answers may include that few people likely live in the Yukon because of its cold climate.

Week 33 Day 2 (page 176)

The Yukon is the top left, with the Northwest Territories to the right of it.

Week 33 Day 3 (page 177)

1. Because of the cold climate, the ground is permanently frozen. This is not good for farming.
2. The Yukon makes money through tourism.
3. Answers may include canoeing, hiking, fishing, skiing, snowboarding, sledding, and dog sledding.

Week 33 Day 4 (page 178)

1. Answers may include that there is grass and no snow.
2. Answers may include that it is cold and icy.
3. Answers may include that people cannot ski or snowboard in the summer, and people cannot canoe or hike in the winter.

Week 34 Day 1 (page 180)

1. the Sierra Nevada Mountains
2. Owens River
3. Answers may include that the aqueduct was built to provide water to Los Angeles.

Week 34 Day 3 (page 182)

1. The aqueduct was built because there was a drought, and the city needed a new water source.
2. Answers may include that it took years to complete, many materials were used, and thousands of people worked on it.
3. Answers should include that the population of Los Angeles increased dramatically.

Week 34 Day 4 (page 183)

1. 2010
2. Answers should be supported by details from the graph.
3. Answers may include taking shorter showers or only watering lawns when needed.

Week 35 Day 1 (page 185)

1. Answers should include that it is in the center of Mexico.
2. United States, Guatemala, and Belize
3. Gulf of Mexico, Caribbean Sea, Pacific Ocean

ANSWER KEY *(cont.)*

Week 35 Day 2 (page 186)

1. The Pacific Ocean, Gulf of Mexico, and Caribbean Sea should be labeled.
2. The United States, Guatemala, and Belize should be labeled.

Week 35 Day 3 (page 187)

1. Answers may include that it shows many people out or the building looks like a museum.
2. Answers may include that people can visit museums, sports centers, or Aztec ruins.
3. The Aztecs were the people who lived in central Mexico for hundreds of years before the Spanish took their land.

Week 35 Day 4 (page 188)

1. Answers may include that they look old, worn, deserted, and that they look like pyramids.
2. Answers may include learning about the region's history or culture.
3. Answers may include that the Aztecs lived in this region long ago.

Week 36 Day 1 (page 190)

1. a circle
2. Answers may include that the railroads made it easier to trade between states.
3. Answers may include that more people were able to migrate to Florida.

Week 36 Day 3 (page 192)

1. to move
2. Answers may include Florida's warm climate or opportunities to find a better life or jobs.
3. Answers should be supported with reasons from the text.

Week 36 Day 4 (page 193)

1. five
2. Answers may include that it helps Florida's economy by bringing money to the state.
3. Answers may include that people want to travel to Florida to visit its theme parks.

NORTH AMERICA

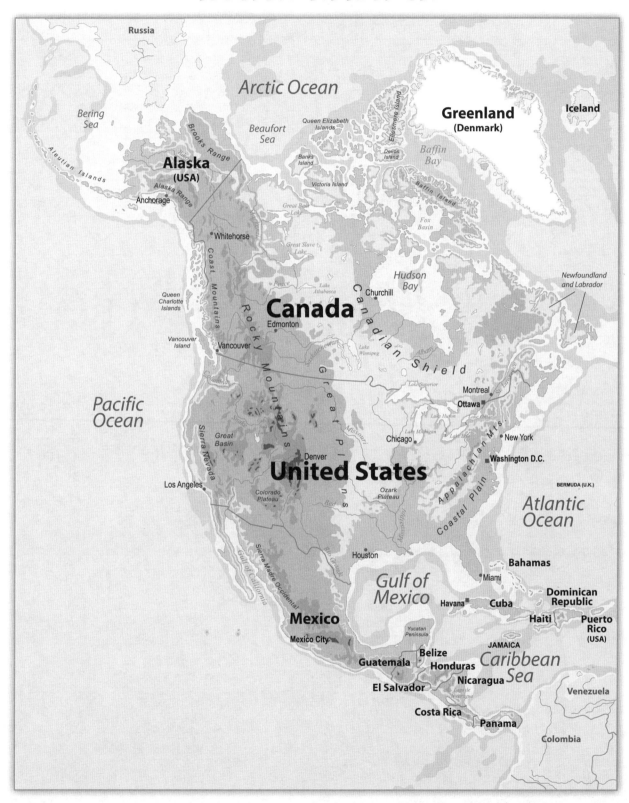

Russia

Arctic Ocean

Bering Sea

Beaufort Sea

Queen Elizabeth Islands

Ellesmere Island

Greenland (Denmark)

Iceland

Brooks Range

Banks Island

Devon Island

Baffin Bay

Alaska (USA)

Alaska Range

Anchorage

Victoria Island

Great Bear Lake

Baffin Island

Fox Basin

Whitehorse

Great Slave Lake

Hudson Bay

Coast Mountains

Peace

Lake Athabasca

Churchill

Newfoundland and Labrador

Queen Charlotte Islands

Canada

Edmonton

Rocky Mountains

Saskatchewan

Lake Winnipeg

Canadian Shield

Vancouver Island

Vancouver

Great Plains

Albany

Lake Superior

St. Lawrence

Montreal

Ottawa

Columbia

Snake

Missouri

Lake Huron

Lake Michigan

Lake Erie

Chicago

New York

Pacific Ocean

Great Basin

Sierra Nevada

Denver

United States

Appalachian Mts.

Washington D.C.

BERMUDA (U.K.)

Los Angeles

Colorado Plateau

Red

Ozark Plateau

Mississippi

Coastal Plain

Atlantic Ocean

Sierra Madre Occidental

Rio Grande

Houston

Gulf of Mexico

Miami

Bahamas

Gulf of California

Havana

Cuba

Dominican Republic

Mexico

Yucatan Peninsula

Haiti

Puerto Rico (USA)

Mexico City

Belize

JAMAICA

Caribbean Sea

Guatemala

Honduras

El Salvador

Nicaragua

Lake Nicaragua

Venezuela

Costa Rica

Panama

Colombia

POLITICAL MAP OF THE UNITED STATES

28624—180 Days of Geography

© Shell Education

PHYSICAL MAP OF THE UNITED STATES

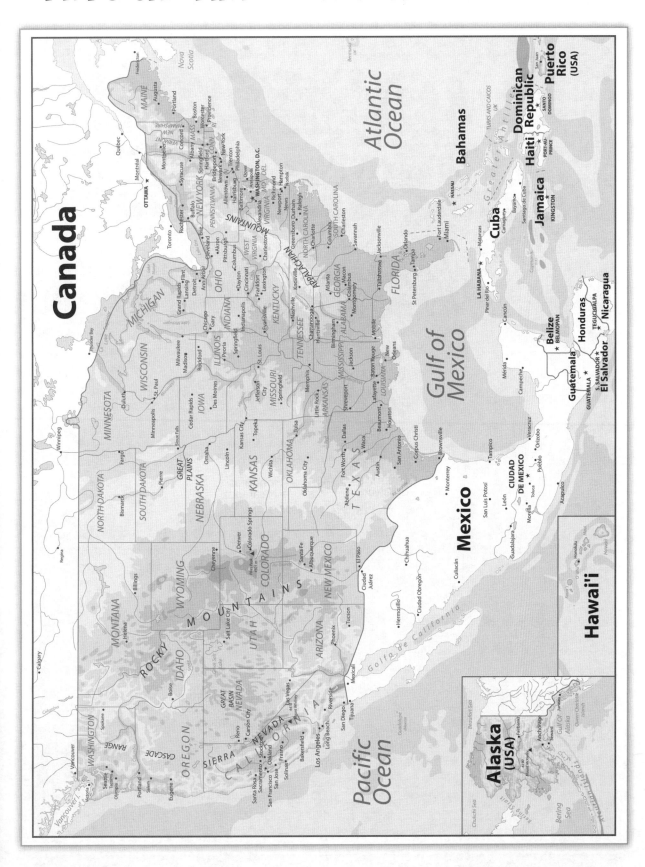

WESTERN HEMISPHERE

EASTERN HEMISPHERE

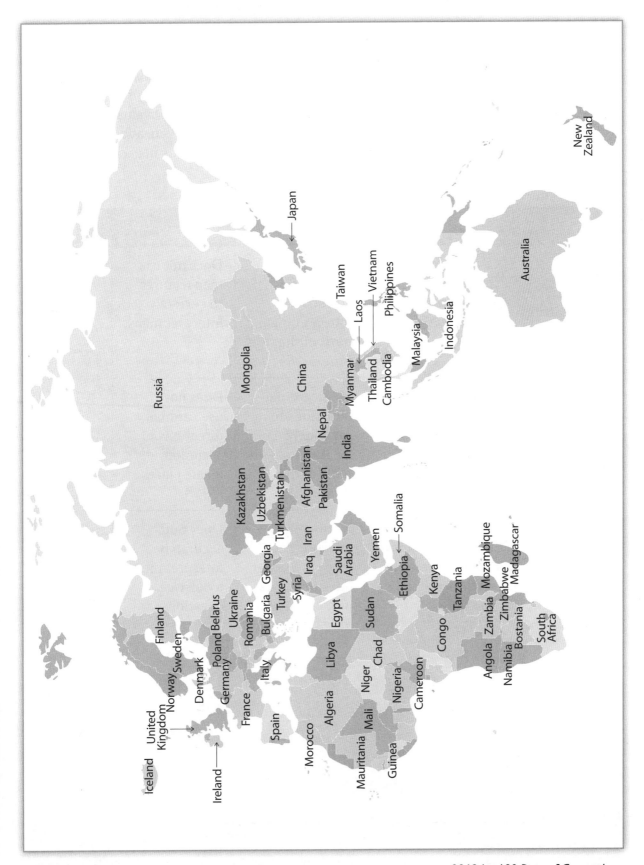

Name: _____   **Date:**_____

MAP SKILLS RUBRIC
DAYS 1 AND 2

Directions: Evaluate students' activity sheets from the first two weeks of instruction. Every five weeks after that, complete this rubric for students' Days 1 and 2 activity sheets. Only one rubric is needed per student. Their work over the five weeks can be evaluated together. Evaluate their work in each category by writing a score in each row. Then, add up their scores, and write the total on the line. Students may earn up to 5 points in each row and up to 15 points total.

Skill	5	3	1	Score
Map Features	Uses map features to correctly interpret maps all or nearly all the time.	Uses map features to correctly interpret maps most of the time.	Does not use map features to correctly interpret maps.	
Using Cardinal Directions	Uses cardinal directions to accurately locate places all or nearly all the time.	Uses cardinal directions to accurately locate places most of the time.	Does not use cardinal directions to accurately locate places.	
Interpreting Maps	Accurately interprets maps to answer questions all or nearly all the time.	Accurately interprets maps to answer questions most of the time.	Does not accurately interpret maps to answer questions.	

Total Points: _____

Name: _____ **Date:** _____

APPLYING INFORMATION
AND DATA RUBRIC
DAYS 3 AND 4

Directions: Complete this rubric every five weeks to evaluate students' Day 3 and Day 4 activity sheets. Only one rubric is needed per student. Their work over the five weeks can be evaluated together. Evaluate their work in each category by writing a score in each row. Then, add up their scores, and write the total on the line. Students may earn up to 5 points in each row and up to 15 points total. **Note:** Weeks 1 and 2 are map skills only and will not be evaluated here.

Skill	5	3	1	Score
Interpreting Text	Correctly interprets texts to answer questions all or nearly all the time.	Correctly interprets texts to answer questions most of the time.	Does not correctly interpret texts to answer questions.	
Interpreting Data	Correctly interprets data to answer questions all or nearly all the time.	Correctly interprets data to answer questions most of the time.	Does not correctly interpret data to answer questions.	
Applying Information	Applies new information and data to known information about locations or regions all or nearly all the time.	Applies new information and data to known information about locations or regions most of the time.	Does not apply new information and data to known information about locations or regions.	

Total Points: _____

Name: _____ **Date:** _____

MAKING CONNECTIONS RUBRIC
DAY 5

Directions: Complete this rubric every five weeks to evaluate students' Day 5 activity sheets. Only one rubric is needed per student. Their work over the five weeks can be evaluated together. Evaluate their work in each category by writing a score in each row. Then, add up their scores, and write the total on the line. Students may earn up to 5 points in each row and up to 15 points total. **Note:** Weeks 1 and 2 are map skills only and will not be evaluated here.

Skill	5	3	1	Score
Comparing One's Community	Makes meaningful comparisons of one's own home or community to others all or nearly all the time.	Makes meaningful comparisons of one's own home or community to others most of the time.	Does not make meaningful comparisons of one's own home or community to others.	
Comparing One's Life	Makes meaningful comparisons of one's daily life to those in other locations or regions all or nearly all the time.	Makes meaningful comparisons of one's daily life to those in other locations or regions most of the time.	Does not makes meaningful comparisons of one's daily life to those in other locations or regions.	
Makes Connections	Uses information about other locations or regions to make meaningful connections about life there all or nearly all the time.	Uses information about other locations or regions to make meaningful connections about life there most of the time.	Does not use information about other locations or regions to make meaningful connections about life there.	

Total Points: _____

MAP SKILLS ANALYSIS

Directions: Record each student's rubric scores (page 210) in the appropriate columns. Add the totals, and record the sums in the Total Scores column. Record the average class score in the last row. You can view: (1) which students are not understanding map skills and (2) how students progress throughout the school year.

Student Name	Week 2	Week 7	Week 12	Week 17	Week 22	Week 27	Week 32	Week 36	Total Scores
Average Classroom Score									

APPLYING INFORMATION
AND DATA ANALYSIS

Directions: Record each student's rubric scores (page 211) in the appropriate columns. Add the totals, and record the sums in the Total Scores column. Record the average class score in the last row. You can view: (1) which students are not understanding how to analyze information and data and (2) how students progress throughout the school year.

Student Name	Week 7	Week 12	Week 17	Week 22	Week 27	Week 32	Week 36	Total Scores
Average Classroom Score								

MAKING CONNECTIONS ANALYSIS

Directions: Record each student's rubric scores (page 212) in the appropriate columns. Add the totals, and record the sums in the Total Scores column. Record the average class score in the last row. You can view: (1) which students are not understanding how to make connections to geography and (2) how students progress throughout the school year.

Student Name	Week 7	Week 12	Week 17	Week 22	Week 27	Week 32	Week 36	Total Scores
Average Classroom Score								

DIGITAL RESOURCES

To access the digital resources, go to this website and enter the following code: 51297342
www.teachercreatedmaterials.com/administrators/download-files/

Rubrics

Resource	Filename
Map Skills Rubric	skillsrubric.pdf
Applying Information and Data Rubric	datarubric.pdf
Making Connections Rubric	connectrubric.pdf

Item Analysis Sheets

Resource	Filename
Map Skills Analysis	skillsanalysis.pdf skillsanalysis.docx skillsanalysis.xlsx
Applying Information and Data Analysis	dataanalysis.pdf dataanalysis.docx dataanalysis.xlsx
Making Connections Analysis	connectanalysis.pdf connectanalysis.docx connectanalysis.xlsx

Standards

Resource	Filename
Standards Charts	standards.pdf